COAL AND ITS TOPOGRAPHY.

MANUAL

OF

COAL AND ITS TOPOGRAPHY.

ILLUSTRATED BY ORIGINAL DRAWINGS,

CHIEFLY OF

FACTS IN THE GEOLOGY OF THE APPALACHIAN REGION OF THE UNITED STATES OF NORTH AMERICA.

BY

J. P. LESLEY,

TOPOGRAPHICAL GEOLOGIST.

PHILADELPHIA:

J. B. LIPPINCOTT AND CO.

1856.

PREFACE.

The author has planted this sapling for the future shade and ornament of his own office, but trusts that it may prove useful also, and perhaps agreeable, to the public highway. To the public, therefore, he offers no apology. But his brethren in science will in courtesy consider how hard a thing it is to treat a large subject in few words; to reduce to reasonable limits a theme of infinite details, which has been pondered long, postponed from year to year, and finally accomplished amid the multiplying avocations of a professional life. To them the only merit it can claim will seem but small,—the presentation of a few results of original research and the expression of independent opinions upon matters of common interest.

In explanation, it is necessary only to add, that the sketches were all made upon the ground and drawn on wood by the author, to preserve with accuracy those points of interest to his own mind which he desires to illustrate in the text, and upon this their whole value as illustrations of truth depends; as works of art they are nothing. In three instances (Figs. 7, 17, 18) he has copied drawings which have been given to the public by others; in one other instance (§ 31) he has, after some hesitation, given data transmitted through a private letter of an old date, without previously obtaining the permission of its author. With these exceptions, he believes his book to be free from actual plagiarisms. It must happen, however, that out of a mass of materials brought together under every variety of circumstances, through nearly twenty years of observation, even the small selection here made must contain many which are due to the labors of others, and bestowed in friendship or business upon the author. He finds it of course impossible to do justice to this interchange of thought and work; but wherever important contributions to science could be identified with those who made them, he has done it, and for the rest can only thank friends of the past and present time in general but respectful and affectionate terms.

To two of these, now in distant parts of the world, men of infinite scope and love in science, poets by Divine right, pure-hearted, true to every duty, to whom, as master in youth and friend in middle age, the author has owed what it would be presumption to attempt in words, and American science what it has never yet had the opportunity to acknowledge—

TO

JAMES D. WHELPLEY

AND

ANDREW A. HENDERSON,

This Manual

IS MOST AFFECTIONATELY DEDICATED

BY

THE AUTHOR.

CONTENTS.

CHAPTER I.

THE GENERAL RELATIONSHIPS OF COAL.

CHAPTER II.

THE PLACE OF COAL AMONG THE FORMATIONS.

CHAPTER III.

TOPOGRAPHY AS A SCIENCE.

CHAPTER IV.

TOPOGRAPHY AS AN ART.

APPENDIX.

COAL.

CHAPTER I.

COAL—ITS POSITION IN GENERAL.

1. **Coal** is never found issuing in veins from the interior of the planet, like gold and silver; nor filling irregular cross crevices in limestone, like lead; nor spread abroad in lakes of hardened lava, like basalt and greenstone; nor embedded in clay, crystallizing upward from the walls and bottoms of deep wide fissures, as bunches of grapes, or in bundles of pipes, like the hæmatite iron ores; nor lying exposed upon the surface in blocks, like native copper, or meteoric iron; but always as a thin sheet or stratum, extending through the hills as far as the hills extend, and inclosed between similar sheets of other kinds of rock.

2. Compared with the sum of all the rocks, in any given mountain, the **thickness** of even the largest coal-beds is insignificantly small, nor does any known mountain contain a sufficient number of such beds to justify the expression so frequently employed by ignorant or interested persons, "it is a solid mass of coal." There are, no doubt, few native business men of Philadelphia, who cannot remember the panic occasioned by the news that the miners had reached the bottom of the Mauch Chunk Summit Mine; after having benched down through seventy feet of nearly solid coal, opening to the sky in a vast quarry, what had been confidently regarded as but the vertex

2

of a solid mountain of the mineral, the depth, and therefore the contents of which were quite incalculable. Lehigh Coal and Navigation Stock fell twenty per cent.; for the same misconception of the general position of the mineral underground, which had bestowed upon it such a marvellous depth, blinded the public to the fact of its much more marvellous lateral extent. Men were terrified to learn that the Mauch Chunk Mountain was not a solid mass of coal—but had yet to learn, that one-sixth or eighth of the United States was underlaid by beds of it. A mountain may be justly called, in a certain sense, a solid mass of gold, for its whole body may be penetrated by and made up of auriferous quartz veins. The same may be said of iron, for in Missouri, and elsewhere, there are iron mountains, solid dykes of metal. But the strata of coal seldom exceed a yard or two in thickness, inclosed between hundreds of feet of other rock; and where in a few favored spots, like the Anthracite region of Pennsylvania and the South of France, mammoth beds of ten and twelve fathoms are worked, upon examination they turn out to be double, treble and quadruple beds, a merely local approximation of several distinct and elsewhere often distant sheets of coal. The seams of slate and rock around the walls of the Mauch Chunk Summit Mine should have prepared us for the appearance of the bottom slate of all.

3. Nothing is more surprising than the vast expanse of even the thinnest of these sheets of coal. The original deposit of carbonaceous matter seems to have been in every instance almost co-extensive with the lake or sea in which it was laid down. Nor can it be considered as satisfactorily explained by any of the hypotheses as yet advanced, how an inch or two of vegetable mould could preserve its sea-level so uniformly over such wide areas as the geographical study of our coal shows plainly that it did. Whatever the future limits to this fact may prove to be, they leave us at liberty to lay down the rule that—Each sheet of coal extends for itself and by itself as far as the mountain does in which it lies, never branching nor fork-

ing nor rolled together, but passing through the mountain from side to side, from end to end, or cleaving down through it from summit to base, from end to end, and commonly in this latter case passing under an adjoining valley and reascending through the length and breadth of a mountain on the other side. All coal beds do this independent of each other while conforming to each other's motions, but never coming into conflict or even actual contact, however near—except in faults, as will be shown hereafter.

4. **Intervals.**—Each stratum of coal lies between other similar strata of flag or freestone, conglomerate (or puddingstone), mudrock (shales or slates), carbonate of lime (or limestone) and carbonate of iron (or ironstone); the whole forming a series, like sheets in a ream of paper; arranged in no discoverable *rational* order, and apparently following each other in no fixed order, but indiscriminately alternating; for the coal will be covered sometimes with a stratum of sandstone, sometimes with shale or slate, and sometimes with limestone; although it very rarely happens that either limestone or iron ore in a stratified form comes into actual contact with it; however close they may be to it, they are commonly separated from it by some inches or feet of common rock, and usually this rock is slate. It is nevertheless true that all these layers, when once their geological order of superposition has been made out for one locality, can be found again related to each other in the same general order of deposition at all neighboring localities, and sometimes even at the distance of many miles from the place where the section was first constructed.

5. *Coal and* **Limestone.**—An opinion prevails in some districts that wherever limestone exists coal cannot be far away. Nowhere is this a safe and proper statement; and over thousands of square miles of the United States, both along the seaboard and inland, along the Great Valley of Virginia and Pennsylvania, and on the plains of Kentucky and Tennessee, no statement of the fact could be less true. The greatest developments of limestone on the crust of the earth have had no

connection whatever with the great deposits of coal, but have either preceded or followed them at long intervals of time. Lesser limestone strata, however, do occur among the rocks which include the coal, as we might reasonably expect; for the ocean seems never to have been without the power of precipitating lime, and seldom to have allowed any long interval of other work to elapse without returning to recreate and literally rest itself for a longer or shorter space with this. The converse of the statement rejected above is, however, with but few exceptions, strictly true; that wherever coal beds exist there limestone beds are to be found; a few feet above or underneath them ; spreading as widely and as evenly as they ; and maintaining with great equalness the intervals between. These limestone layers vary infinitely in character and size, from a few inches to scores of feet in thickness, and from pure hard blue rock, magnesian lime, cement layers, or a double carbonate of iron and lime, to the softest marls or clay slates containing here and there a single nodule. And the integrity of these deposits is wonderfully well preserved over vast areas; for the same bed on which are opened inexhaustible quarries of limestone in a dozen counties will continue to range over still other hundreds of square miles after it has diminished and impoverished itself to a thin and worthless seam.

In Nova Scotia, gypsum, or sulphate of lime, is seen to take the place of limestone or carbonate of lime, among the coal, and probably not by virtue of any subsequent change effected by the sulphuric acid set free by the decomposition of iron pyrites in the coal and neighboring rocks, but by virtue of some local deposit of sulphate of lime from volcanic springs in the bed of the carboniferous ocean; for otherwise it is hard to understand the almost universal presence of limestone and absence of gypsum in the coal formation. It is true that the gypsum quarries of Southern Virginia lie against the edges of the coal, but they also there occupy great faults in the crust of the earth which bring up to the surface the lowest rocks known in the ancient series, and offer a very different explanation of the presence of the gypsum.

6. *Coal and* **Iron.**—Still more universally true is it that wherever there is coal there is also iron; either in the bed itself, or in the rocks above it or below; either in the form of nodules of carbonate of iron, or in cubic crystals of sulphuret of iron scattered or conglobed; or as fine disks between the faces of the coal; or spread in thin wide plates of carbonate, splitting into masses hundreds of pounds in weight, ranging hundreds of yards or even miles in extent, and sometimes passing by insensible gradation into beds of carbonate of lime; the local deposit of iron giving place all round the locality to a general deposit of lime. Throughout the coal measures, ores of iron are always carbonate except when in contact with or included in the coal, in which case they are always sulphuret. Whether the sulphur of the original vegetation originally combined and was precipitated with the iron in the coal, or whether in the form of free sulphuric acid, it has in course of time transformed an original precipitate of carbonate of iron into sulphuret, is a question yet. The latter is more probable, although it exactly reverses the process which has been going on from unknown dates in lead and copper mines, wherein the ore was an original sulphuret, but has been leached down to water level, and that far converted by rain-water agency into the oxide and the carbonate. But whatever may be our explanation of the coming of the sulphur, its presence is a serious evil both to the iron and the coal; the former it makes worthless for the furnace, the latter it injures for the forge. Nothing is worse for the reputation of a mine than its being reputed to abound in crystals, or nodules of iron pyrites; and more than one bed has maintained its good name by being carefully picked. The coals of the far West take a low rank, because of the amount of this injurious deposit in them. Hundreds of tons of balls and cakes of pyrites have been sold to the vitriol factories in past years from the openings opposite St. Louis. This is the only purpose to which the pyrites has been applied,

and gives it but local and precarious value. With this trifling exception, it is worse than worthless. Coal beds differ greatly and characteristically among themselves in their gross percentage of this sulphurous alloy, and rank accordingly. But even a nicer distinction still is made between the several plies or seams of coal in a single bed as to which one practically carries this offal of the mine, for it is rarely distributed in equal proportions through the coal, but on the contrary affects some one bench more than the others; sometimes the lowest, sometimes the highest in the gangway. If the coal bed be a large one, such a sulphur bench is either left unmined or rejected at the mine's mouth with the slaty and rocky portions of the bed. It is to the presence of sulphur that the noisome smell of coal gas is chiefly due, and the astringent and disgusting taste of springs that issue from the hill-sides that contain coal. Free sulphuric acid is always present in these, with the sulphate of iron, depositing a yellow slime in the channels of the brooks that flow from them. The carbonate of iron is as valuable as the sulphuret is worthless. When fused alone, it yields one of the best metals in the world, as is now shown at the Cambria Works in Pennsylvania. But a mixture of the sulphuret, either in the ore heap or in the coal, ruins its iron. In itself it is of every quality, clayey, sandy, or pure, fine grained or coarse, blue, gray, or chocolate tinted, breaking with a poor, loose sandstone fracture, or with a rich, fine, sharp, conchoidal, jaspery surface, showing the absence of silex, the low percentage of argil, and the superiority of its iron. While all the furnaces of Eastern and Middle Pennsylvania run upon the oxide hæmatite ores of iron, all those of Western Pennsylvania run upon these coal measure carbonate ores. But in the majority of instances they have been built in the neighborhood of very insufficient beds, and a large proportion of those originally erected blew out for want of ore, and form picturesque ruins in secluded glens among the mountains, and on the banks of the principal affluent waters of the Monongahela and Alleghany. Although the precipitation of carbonate of iron went on during the whole coal era, and the most barren shales and sandstones abound in its concretions, yet this very abundance of deposit has gone against the miner, for it has distributed his ore beyond his reach. He sees on all sides of him the precious metal—can pick it out from every stratum in the hill before him—can find vast plates of it in the beds of torrents, at the foot of falls, and on the slopes of broken rock below the cliffs, but he is never sure of his finest exposure, his most promising outcrop, that it will not fail him before his gangway has gone in ten yards, his plates breaking up into balls, and the nodules dwindling and separating in a mass of shale. This treachery of the *beds* of carbonate of iron (the ore is good enough), rather than any want of skill or capital, or tariff, has been the secret cause of the periodical and almost universal failure of iron-making in Western Pennsylvania from the beginning until now. And only when built upon well-known and rare deposits, such as the burr-stone ore of Clarion and Franklin Counties, or local exhibitions *first well proven*, as at Johnstown, Cambria County, have furnaces justified expectation. That there is incalculable wealth in iron of this kind throughout the United States no one can doubt, but in each individual case the risk of failure in the ore is very great, and must be met by careful previous exploration.

In Scotland, the carbonate of iron attains such a degree of purity as to excel all other forms of this ore. In the ordinary ores, 35 per cent. of protoxide of iron, with 30 per cent. of carbonic acid to be driven off in fusing, and the remaining 35 per cent. of lime, magnesia, sand, clay, sulphur, manganese, and water, to be lost in slag, is considered a fair quality of ore; but in the best *blackband*, the earthy slag amounts to almost nothing, 5 or 6 per cent., and the protoxide of iron amounts to 50—55 per cent. of the whole. Besides which there is a percentage of *free coal* in some of it which helps to smelt it, while the iron that it makes is the softest and most fusible known in trade. An important drawback to these advantages, however, exists in its coldshort, or brittle quality, which no treatment can cure, and by virtue of which, while it holds the first rank in castings, it is worthless where strength is requisite, and ruins other iron with which it may be mixed for almost all the purposes of the machine shop and the railroad. We may therefore convert our repeated disappointments at its published discoveries in this country into hearty congratulations that it scarcely exists in our coal measures; for while it has created enormous personal fortunes, and stimulated for a time the local iron trade of Scotland and England, it has deteriorated iron on both sides of the Atlantic: whereas the time has fully come for the successful and profitable treatment of the common carbonate at innumerable points, either pure or mixed with the bog oxide, or with the hæmatites, or with the fossil ore, fused with the raw semi-anthracites of Shamokin, Broad Top, and Cumberland, or the coked bituminous coals of the great West, and fluxed with limestone out of the same hill-side from which both the coal and the iron are obtained.

So much of the iron manufactured in Great Britain, especially in Scotland, is made from the carboniferous ores, that the following sums of the three ingredients used in the manufacture officially reported, may be employed to obtain a practical average of their relative proportion, such as is frequently inquired for. We are told (see *Hewett's Statistics*, p. 12) that in 1854 the consumption of *Coal* was 20,146,000 tons; of *Iron Ore* 12,346,000; of *Limestone* 2,450,000; and the product Iron 3,585,906, which gives the average relation:—

Coal : Ore : Lime : Pig : : 5.6 : 3.4 : 0.7 : 1.

One of the most curious facts connected with the presence of disseminated iron in the American coals is its superabundance in the upper beds, and thence the distinction of a lower white ash system and an upper red ash system, with some intermediate gray ash beds.

7. *Coal and* **Fireclay.**—Under every coal bed is spread a sheet of fireclay. It is a universal rule, although exceptions do occur. This clay is whitish, sometimes very white, solid and tough; often with an unctuous fuller's earth feel, but

oftener with a fine grit between the fingers or the teeth, enough to show the presence of silex in minute subdivision; breaking irregularly sharp and exposing dark cross lines or bars, the rootlets of an ancient water tree. Coal beds are known that have a sheet of fireclay in their midst (Fig. 1); such beds are thus resolvable into double beds; and the parts will always be found, if followed far enough, to separate and remain apart as two coal beds, each with its own floor of fireclay. The fireclay

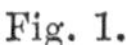

Fig. 1.

deposit under the first great coal bed of the series is now known to be of the most irregular kind. It varies from one foot to forty feet in thickness, and appears to have filled up a most uneven ocean bottom, full of hills and hollows. In fact the presence of this clay beneath the coal not only establishes the fact that each coal bed was created during a time of quietness, and destroyed by the following tumultuary inroads of rock material; but also that no coal was possible until the vestiges of previous commotion in the surface had been erased and the dead level quite or very nearly restored by the abundant precipitation of mingled gravel, sand, and mud, concluding with this clay. In it the incipient vegetation struck its roots, and the clay is often a matted mass of rootlets. Nevertheless the clay is not an absolute necessity, for it is sometimes, though rarely, replaced by sand. Sandstone sometimes, also, parts a coal bed, commencing with an inch, and ending with several feet of thickness; the sandstone parting of the Cook bed in Broad Top, after becoming three and a half feet thick, splits in its turn and receives a knife edge of coal which constitutes a third and middle bed between the other two. On the other hand, fireclay sometimes (as in the case of this very bed) forms a ponderous covering for the coal.

8. **Blackslate.**—This is often mistaken for coal, or for the "flower" of coal. It is a deposit of mud highly charged with vegetable matter, bitumen, or carbon, but not enough so to make it burn. When thrown into a lively fire it becomes red hot like the coals, and loses what little carbon it contains, but in the end clogs the grate like a stone, and cannot keep alive of itself. It can always be distinguished from true coal by its splitting into flakes, especially when heated. Coal never does this. Those parts of a coal bed which show this tendency, and are made up of alternate layers of blackslate and coal as thin as paper, the whole mass hardened by the presence of sand and iron, the miners call "bony coal." It is no peculiarity of one bed, but exists more or less in all; nor of a locality, but is everywhere one phase of the double deposit of sandy mud, and the vegetation among which it was conveyed by currents. Some beds, it is true, are remarkably free from slaty bands, and for a few yards a gangway may proceed to take out pure coal alone; but knife edges of slate are sure to come in; while, in fact, many beds are called coal beds only by courtesy, or by virtue of their coal elsewhere, for they exist as an entire deposit of blackslate. Both these extremes, however, are rare. As the floor of most beds is a fireclay, so the roof is commonly blackslate, showing a failure of vegetable matter near the close of the time in which the bed was laid down. There are beds of blackslate of a date much earlier than the true coal measures, around Lake Erie and on the Hudson River, and in the central valleys of Pennsylvania and Virginia, in exploring which fortunes have unhappily been squandered, the amount of interleaved coal never being sufficient to repay mining. So in the very heart of the great coal formations there are barren regions wherein beds of blackslate offer a perpetual fallacious inducement to unsuccessful adventure. Even where the true coal abounds they are not always distinguishable at the surface from such

carbonaceous slates, inasmuch as the bony portions of a coal bed are the least destructible by atmospheric agencies, and therefore govern its appearance along its outcrop. But the principal mistake to which the searcher after coal is now subjected consists in mistaking the more compact, fine, and homogeneous varieties of blackslate for cannel coal.

9. **Cannel Coal** is carbon, nearly pure from clay and sand; while blackslate is characterized by its excess of clay. When the blackslate is charged with from thirty to fifty per cent. of carbonaceous matter, it greatly resembles, and has been repeatedly presented for true cannel coal. The uncommon qualities of cannel, together with its acknowledged scarcity, has stimulated the monopolizing speculation spirit of the day to the trial of its representatives whenever they appear, and announcements have repeatedly been made of the discovery of enormous beds of it, which were, in fact, but beds of compact blackslate. Apart from a chemical analysis, three convenient tests may be applied by any one in doubt; the weight of blackslate is plainly greater than that of cannel coal; a splinter of it will not light and blaze like one of cannel coal, but merely glow in the flame where it is held; and thirdly, a heap of it on the ground will not make a fire unless assisted by wood or coal, whereas a single lump of cannel laid upon an ember will be entirely consumed. In what other respects cannel differs from the high per cent. dry bituminous coals with which it is associated (usually as superior and inferior layers of the same bed), is not yet well understood. The interior molecular structure of compact cannel and of columnar bituminous coal, must be very different. The former almost always exhibits the conchoidal fracture peculiar to clay rocks; the latter never. And in some varieties, as at Peytona in Western Virginia and elsewhere, there is a bird's-eye structure peculiar to the cannel and unexplained, which has, probably, never been seen in blackslate; an approach to it is occasionally made in the harder kinds of anthracite. The essential family relationship of all these deposits, however, is undisputed. Cannel, Glance, Splint, Bony Coal, Blackslate,

are but so many gradations in a mixed and ever varying proportionate deposit of carbon and sandy clay of an original vegetation, and the slow current ooze of the shallow ocean in which it grew, or into which its leaves and twigs were floated in a condition of fine mechanical subdivision and solution. Where the vegetation was weak, and the mud in excess, blackslate was the deposit; but, where the vegetation was densest, and offered the most effectual barrier to the current and the influx of its mud, cannel coal was formed. This explains the local nature of this valuable mineral, and the fact of its gradual disappearance in all directions from a centre of maximum thickness, or the conversion of its layers into bituminous coal or blackslate.

10. **Anthracite** is not an original variety of coal like the above, but a subsequent modification of the same beds, which, in other parts of the region, remain bituminous. Anthracite beds, therefore, are not separate deposits in another sea, nor coal measures of another era, nor interpolations among bituminous coals; but the bituminous beds themselves, altered by subterranean vapors and transmitted volcanic heat into a natural coke from which the volatile bituminous oils and gases have been driven off, leaving the solid carbon with the mud, sand, sulphur, and iron behind, recrystallized in a more compact and durable form, more difficult to mine, but better able to bear carriage, and containing a larger quantity of fuel in a smaller bulk than before. The anthracite character of the coal beds is never found but in the most disturbed regions, for instance, in South Wales, in the Alps, and along the seaboard of the United States. Here the transition of anthracite into bituminous is not so easily studied, because the coal formation has been destroyed to such an extent that only outlying patches of it are left. But in Wales the same bed may be traced from where it is bituminous to where it is anthracite, through all the gradations of the change. Whatever, therefore, has been said of bituminous coal beds is equally true of them in regions where they yield nothing but anthracite. The rocks of the intervals, however,

are, in the latter case, also more or less baked and changed, hardened, and cleft in crystal planes, by the same agency that has metamorphosed the coal.

By what has gone before the author is far from ignoring the critical distinctions which the mineralogist and microscopist make between the varieties of coal, nor the suggestions which botanists advance towards a final resolution of these inherent difficulties involved in the question : What is coal ? The opposite results of the two recent lawsuits in England and in Prussia, based upon the character of the Torbanehill variety of the mineral, results so illustrative of the opposite genius of the two nations, show both the diversity of feeling on the subject among men of science, and the technical character of the discussion. The English courts decided that whatever looked and burnt like coal, was mined and used as coal, and lay in the usual form, position, and general relationship of coal, was *practically* nothing else than coal—a practical English conclusion. The German courts on the contrary ruled, that what the microscope and crucible separated, no gross mechanical convenience should be permitted to confound—that what has not the essential ultimate traits of coal has no right to its name—in fine, that the interests of the gas company and of the custom-house are not to be thought of where the interests of a true scientific nomenclature and the final adjustment of human ideas are in question—a true ideal and Teutonic conclusion. Among Englishmen themselves there is the same dissension. Some maintain that no distinction except one of degree can be made clear among the members of the series from Parrot and Cannel coal to blackslate or carbonaceous clay. But it is impossible to avoid the claims of a multitude of observations made by another class of observers to be admitted into the final discussion of the history and nature of coal deposits of all kinds. A cone-bearing tree has been seen rooted erect in one coal bed, penetrating upwards through a system of rocks, through a second coal bed and broken off in a second system of still higher rocks. Two coal beds and many layers of sand and mud therefore were deposited around this standing *fir* tree. What was the vegetation out of the more rapid decay of which these two coal beds were made? It is easy to believe that it was a soft spongy marsh fern or peat bog vegetation. But the spiral tubes of the *mosses* are not discoverable under the microscope in most coals; the circular holes left by the decay of straight tubes such as characterize certain *ferns*, are. The innumerable sigillaria trunks and stigmaria roots above and below the coal make it almost certain that the mass of the coal is a decay of ferns. On the other hand certain coals exhibit disks with centre points, and these are characteristic of the cone-bearing woods. We know that the pine or fir tribe abound with resinous gums which, when fossilized would furnish the strata with their bitumen. The distinction of dry and fat coals, between bituminous fossil clays and compact cannels, and perhaps between certain bituminous and certain other anthracite beds is to be explained by an overplus of resinous pines in the ancient coal forest, either in the swamp itself or swept into it from the uplands. It is certain that of two beds equally disturbed or equally undisturbed, and laying not many yards the one above the other, one will contain a higher average of bitumen than the other; and in the remarkable instance of the litigated Torbanehill bed the whole stratum was charged with bitumen to the almost total exclusion of the fibrous forms of vegetable carbon. And no doubt the varieties of cannel coal are to be explained in a like manner. We have therefore coal with much or little clay, with much or little bitumen; clay with coal without bitumen, and with bitumen with very little coal; and all these varieties either as they were deposited or subsequently changed by volcanic heat, debituminized and semi-crystallized. The immense range of variety thus obtained and the insensible gradations by which they must pass into each other may be better imagined than described. It will be the serious business of geologists to discuss and describe all these varieties until they are understood and published to the business world, for every change in the character of coal only provides a new and more efficient agent to some branch of the arts which has been waiting for it to complete its own development.

11. **Coke** is the solid carbon and ash in coal, obtained by driving off the water, the hydrogen, the sulphur, and any other volatile matters which the coal may originally contain. These volatile matters weigh nearly, or quite as much as the solid matters, but fly off without changing greatly the solid form. Hence a bushel of coal, before coking, remains a bushel after coking, but weighs only half as much; is vesicular, or porous, splintery, crystalline, and sonorous.

The bulk of coke varies with the method of obtaining it; a quick fire, under heavy pressure, makes a hard, firm, heavy coke, silvery white, and ringing when struck. A slow, smouldering fire, makes a light spongy coke. By firing slowly at first, with a moist heat, and afterwards, when the sulphur is gone, firing up rapidly, both purity and weight are attained. Overman recommends coking in rows (p. 121), or heaps, a hundred feet long, seven or eight wide, and three feet high, in a level yard, surrounded by a ditch always full of water. Coarse brick chimneys, with holes, are built along the centre of the row, coarse coal piled round them, draft channels left along the ground, and the whole covered with coke dust. In a few hours the whole will be ignited, and a few air-holes, made in the coverings, will allow the sulphurous acid and other fumes to escape. Tapered posts stuck in the ground every seven or eight feet, and withdrawn after the coarse coal has been piled round them, will serve instead of chimneys. The firing takes place when the row is—say 20 feet long; before the row is finished, and fired at the far end, the coke may be drawn away from the beginning, and the process be perpetual. When the white fumes of carburetted hydrogen cease, the fire must be closely covered up, and the mass left to cool. In coking high bituminous coals, the fire should spread through the whole mass before the covering is put on, otherwise the swelling of the softened lumps will choke out the fire; for the same reason, the more bituminous the coal the larger should be the pieces first laid down, and care should be taken to stand them on their *ends*.

The principal object of coking is to free the coal from *sulphur*, for smelting iron; most coals being sulphurous, and sulphur ruining the malleability and tenacity of iron. The object is effected by piling the coal on moist earth, or filling it into ovens with a sand floor kept moist and firing it slowly, so that the sulphur may escape first. If a dry high heat be applied at once to the mass, the water, hydrogen, and bituminous compounds, will be driven off first, leaving the sulphur indissolubly combined with the carbon. Such coke, when used in the furnace, is as bad for iron as the sulphurous coal of which it was made. The sprinkling of water upon red hot coke which has not lost all its sulphur, will prove its existence by the smell of rotten eggs, but will not rectify the coke.

The weight of coke of course varies greatly, because it depends upon the amount of solid carbon and the amount of earth in the coal. As 100 parts of coal may contain from 6 up to 60 parts of bituminous or volatile mineral, from 30 up to 90 parts of solid carbon, from 1 up to 20 parts of sulphur, from 5 to 50 parts of silex, alumina, potash, magnesia, lime, and iron, the residuum of coke obtained must be as infinitely various as the possible combinations of the two variable proportions of carbon and earth—even supposing the process of coking to be perfect. But as no coal can be called first quality which has more than 10 per cent. ash, an anthracite which has, say 10 per cent. bitumen, should yield 80 lbs. coke to the hundred, while a bituminous coal with say 45 per cent. of volatile matter should yield but 45 lbs. of coke to the hundred; if the ash fell to 5 p. c., the coke would rise to 50 lbs.; if the bitumen fell to 40 p. c., the coke would rise again to 55 lbs., &c. This is the theory. In practice, however, the process is not perfect, and the weight of coke is increased actually by the remnant of substances not volatilized by oxygen obtained, and proportionally by a certain percentage of carbon lost by combustion—so that the coke of a hundred lbs. of coal weighs from 90 to 55 lbs. when made in ovens, but only from 80 to 45 lbs. when made in the open air. (See *Johnson's Report to the Navy Department.*)

Overman makes the important assertion that "the coal of the Western States is generally of good quality, particularly the veins lying above the extensive Pittsburg vein. The lower veins do not yield so good an article for the blast furnace." This is a dangerous generalization, made at too early a day upon insufficient data, and must be taken with large allowance. It is *apparently* true that the lower beds are more irregular and more sulphurous. But the percentage of sulphur is known to increase westwardly, and only the lower beds spread over the far west; the beds above the Pittsburg vein do not cross the Ohio; it is impossible to say how sulphurous they were in the far west from which they have since been swept; nor have we yet the means of deciding the relative merits of these upper beds, and those of the lower series where they overlie each other, except in one place along the line of the Pennsylvania canal where the outcrops of the two systems are brought close together, and where the lower coals are quite as bituminous and non-sulphurous as the upper.

12. *The* **Practicable Character** *of a Coal Bed* is soon determined by a few good openings upon its outcrop. It is a general opinion among all classes of people, even the most intelligent, that all mineral exposures, coal included, improve in quantity and quality when followed into the earth, an opinion naturally prompted by the mystery of depth and by the hopes of gain, but none the less fallacious and expensive. A sample of a coal bed can be had at its outcrop as easily and safely as a sample of broadcloth can be cut from the end of a roll. Under ordinary circumstances and for all practical purposes the quality, size, and mining condition of a coal bed can be explored as well in two or three days by a gangway ten or fifteen feet long, as by workings through it for a month or a year. In this matter the word of a miner is never to be taken.

His interest is to mine indifferently through coal, blackslate or rock; and his daily employment fosters in his own mind the same false expectation of improvement in his gangway which proves so seductive to his employer, the owner of the land. A poor bed, when once fairly seen to be poor, ought never to be pursued unless the property contains no better, or unless the bed itself can be opened nowhere else. The only proper method is to open the same bed at numerous places along its outcrop, and from a comparison of these crop openings the actual average character of the bed within can be confidently predicted, and its contents calculated. But then these crop openings must be each one thoroughly made, the very top and the very bottom of the bed ascertained, and the whole thickness pursued far enough under cover for all the soft and weathered materials to turn into solid coal and slate. The custom is the very reverse of this. Shallow superficial holes are hastily dug wherever coal-smut happens to be observed, and left to be filled with water even before the full depth of the coal is reached. Neither top nor bottom is obtained, nor the dip of the bed, nor the amount of slate. Nothing is known after scores of such holes have been dug that was not known before, namely, that coal existed; but how, how much, in what posture and condition, are questions still unanswered after the expenditure of money and time enough for a complete geological survey of the whole property. Incalculable wealth, impatience, anxiety, and delay, would have been spared to multitudes of people had the countless holes and trenches dug in all our mountain ravines been cleanly cut, at an ascending grade, five yards into the hill-side, and the careful examination and measurement of the exposed wall been accepted as a sufficient index of the size and structure of the bed. A coal bed may indeed belie itself at one point or at two where openings are made, but not at a dozen. Veins of lead, of copper, of iron alternately increase and diminish in size, rapidly and unexpectedly. The miner never knows until he strikes the vein what it will be worth, nor how soon the pocket which he has entered may close up between bare walls. Not so with coal. It varies little and seldom disappoints. What it is at one point it is likely to be much the same at another. Even in disturbed anthracite regions where convulsions of the crust, rock faults, crushes, down-throws, have displaced it to any upper or lower level, they do not ordinarily change its form and character, and where they squeeze a portion of its contents forwards or sidewise at one point, they only thereby add to its size at other points. Even intervals of hundreds of miles in which it has undergone an infinite number of slight and perhaps some striking variations, will sometimes present it at the most distant places with strangely identical features, showing how vast and regular has been the law of its deposit. One or two of the lowest members of the series, for instance, are characterized from Northern Pennsylvania to Southern Kentucky, by their tendency to run locally into cannel coal. The great Pittsburg bed is a remarkable instance of this law, covering as it does tens of thousands of square miles, and scarcely varying from a thickness of eight feet, showing itself always a double bed, and yielding everywhere both a superior quality and quantity of coal.

13. **Surface Indications.**—As might be expected from what has been said, the features of the surface have little or nothing to do with the character of the coal beneath it. A coal bed is not necessarily regular and good where it underlies a smooth hill-side or alluvial plain, nor necessarily broken and bad where it crops out beneath a cliff, is cut into by a ravine, or the land above it happens to be rough. If the coal in a gangway degenerates in the neighborhood of a gully, nineteen times in twenty this will have been an accidental coincidence. Miners find it difficult to divorce the two sets of phenomena, those of the interior and those of the surface. In geology, however, it must always be kept in mind that the coal bed was made long before the mountain, and that the subsequent shaping of the mountain neither injured nor improved the beds within it, for their character had been determined forever previously.

Whatever exceptions to this rule exist come under the head of dynamic changes, or crush faults, and will be discussed in their appropriate place. But now, however little the final shaping of the earth's surface may have influenced the character of its subterranean contents, the latter has had great effect upon the former. While, on the one hand, the cutting out of a group of hills, the ploughing down of a valley's branches, may have left the edges of coal beds bare, but must have left the interior of the beds unchanged—may have swept a vast body of coal to an unknown distance and greatly diminished the wealth of a region, but cannot have injured either the quality or the quantity of what is left—it is, on the other hand, equally true that the presence of the softer coal beds and their clays, interleaved among hard sand-rocks and massive conglomerates, must have modified this cutting and shaping process at every stage of its advance. In other words, the internal geology of every region has governed according to the most precise laws this external topography; and as a consequence of this, the topography is always our best guide to the geology. In coal regions this is especially true. Every lineament of the surface indicates a mood in the geology beneath, so that the whole face of the earth, like the countenance of a man, is instinct with living impressions of its past experience.

14. **Terraces.**—The principal surface indication of a coal bed is a coal bench, or terrace. The rule is, that every coal bed makes a terrace along a hill-side and a hollow at its end; for the surface having been cut to its present shape by floods of water, or of water mixed with sand and ice precipitated over it, the softest strata suffered of course most erosion, and the edges of the harder and more massive rocks were left in ridges, hills, and mountains. Coal belonging to the softest rocks has suffered more than the harder or compacter strata which inclose it. However deeply they were cut away, it was always cut away still deeper, and its outcrop left in a crease along the surface. The more massive and refractory the rocks which lie above and below the coal, the sharper and more legible to the

eye of the spectator is this crease which it makes between them, or the terrace by which it separates their outcrops. When the coal and other rocks lie horizontal, successive terraces surround the hills, and ascend the branching valleys in all directions. Perhaps the most remarkable exhibition of this kind is at Fort Hill, in Somerset County, Pennsylvania, near the Turkey Foot, or confluence of Castleman's River with Indian Creek and the Youghiogheny, where a quadrangular truncated pyramid may

Fig. 2.

be seen, so regular in form, and so regularly terraced by successive coal beds, that the excited superstition of the aborigines invested it with supernatural sanctity, and converted its extensive level summit into a necropolis, where skeletons are still turned up at every ploughing. Instances of the same phenomenon abound throughout the western country. When the dip of the coal is steep, and the intervening rocks are sand or puddingstone, the terraces along the side of the mountain are not less striking, but at every gap they may be seen to sink in a row of steep ravines to water level. Perhaps no better place can be referred to to study this than some one of the gaps in the Shamokin Mountain, in Pennsylvania.

Fig. 3.

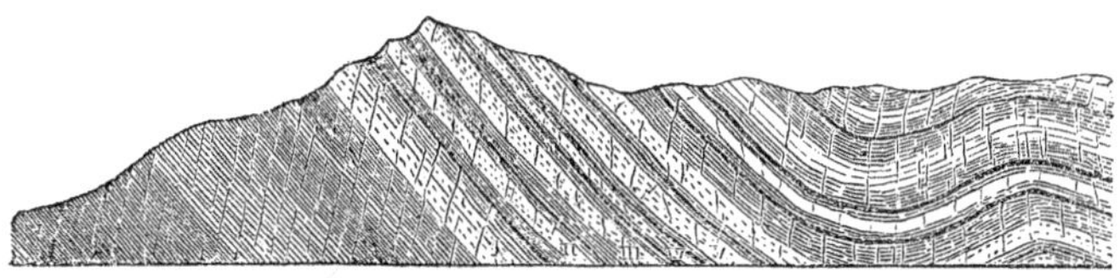

15. **Springs.**—The fireclay which underlies most coal beds presents an impervious curtain to the myriad tiny threads

of rain-water which perpetually follow down all the fissures of the rocks. Turned upon this slippery surface the water flows upon it through the coal bed, or rather between it and the clay, and issues in a line of springs, along the outcrop of the bed. If the coal be pure, the water of these springs will be pure. If the coal, or the slates immediately above it, contain much sulphuret of iron, the water will be charged with iron rust, copperas, vitriol, and alum; wherever it stands, all the colors of the rainbow will glimmer on its surface, and, if the iron be in excess, dome-shaped bogs of spongy brick-red ore will form at the foot of its first slope, although these last properly belong to the drainings of beds of carbonate of iron, and not of coal. A line of *painted springs* along a terrace is the characteristic index of coal. The amount of water which may issue from a spring is in no proportion to the size of the coal bed from which it flows, but is determined by the amount of surface drained, the slope of the hill-side, the character of the upland, the dip of the rocks—especially the thickness of the interval between its water-bearing stratum and the next one above it—the relation of the spring to the surrounding water-level, and other accidents. The largest springs may mark the presence of only an inch or two of coal, while the most important coal beds happen often to be almost dry. As much depends upon the roof of the coal as upon its floor. Sometimes beds have impervious roof-slates, and all their springs will issue then above them. Occasionally coal beds have broken floors, in which case the drainage will keep on downwards until it meets some bed of clay. The dampness or dryness of the mines is decided on the same principle; just as caves in horizontal strata, with flat, unbroken ceilings, are dry and comparatively destitute of stalactites, while caves in uptilted limestone, with ceilings presenting the parallel edges of the layers, between which the rain-water trickles in drops and sheets, are crowded with pillars and pendants, and festooned and tapestried with alabaster—the difference between the Mammoth and other western caves, and Wier's and other Virginian caves.

16. **Black-dirt, Blossom, Smut** or **Crop,** are names given to the lampblack soil into which coal weathers down after the exposure of centuries or millenniums to the action of the air. The edge of every coal bed is in this condition for several feet or yards in from the surface, except where it is washed by running water. Every rock presents its own contribution to the general soil, which must by no means be considered a subsequent deposit on the rocks, nor a product of vegetation, but a decomposition of the edges of the strata effected by the weather of geological ages. Hence soil lies in streaks of sand and clay, or a mixture of the two, according to the run of the different kinds of rocks beneath it; is full of lime on limestone, and full of iron on mica slate; is too stony except for woodland purposes wherever ribs of freestone or puddingstone crop out, but is always seen fenced and farmed where the shales appear. Black-dirt is the soil of coal, and a positive index, therefore, of its presence underneath. It is frequently imitated by the humus of new ground mixed with the charcoal of fresh clearings in new countries, but a single blow of the mattock will distinguish the one as superficial from the other, which grows blacker and thicker as it penetrates the earth. The position of the coal is always *above* the smut, for all soils slide; and, in fact, the whole surface of all hills have been in slow but perpetual movement downward from the beginning, so that in the present day the soil or weathered broken edge of any stratum overlies the strata below it, while it is itself covered by the soil of some stratum above it. On slight slopes this translation of material has gone no great distance, but on slopes of 20° or 30° the smut of a given coal bed has probably been drawn out in a long knife-like wedge, the edge of which is to be seen many yards below its proper place. In certain cases an extraordinary inversion is the consequence; for a coal bed which dips steeply, as many do, not inward towards the centre of the hill, but outward towards the bottom of the valley, will have its outcrop turned over and laid back loopwise down hill, apparently reversing the entire structure of the hill, as shown

in the diagram annexed. In all cases the leading of the black-dirt must be followed up the slope, and the bed opened where it finally enters between solid rocks, whether of slate, or shale, or sandstone; and no coal bed can be said to be opened until all loose and broken stuff has disappeared from the roof and given place to consistent regular layers of slate or other rocks.

Fig. 4.

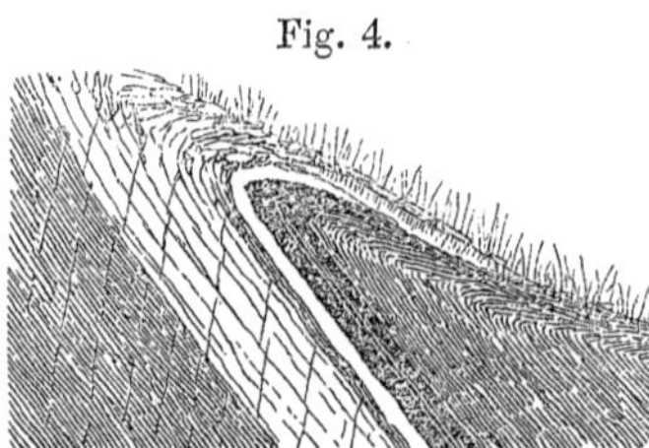

The upturned roots of fallen timber are the usual discoverers of coal. These are reported by hunters, who easily detect fine cubes of coal, fragments of blackslate, or traces of black-dirt, among the upthrown flags and shales. In the springs above described, and in the rills that flow from them, a careful examination will detect similar small crystals of coal and flakes of jet blackslate, embedded in a fine whitish or mottled and tenacious clay. In larger brooks, which cut alternately against one hill-side and the other, baring to view, frequently, the solid face of a coal bed, or flowing over its upper surface as a floor, and tearing it away piecemeal, hand specimens may be met with far down the streams, and these will lead the explorer

Fig. 5.

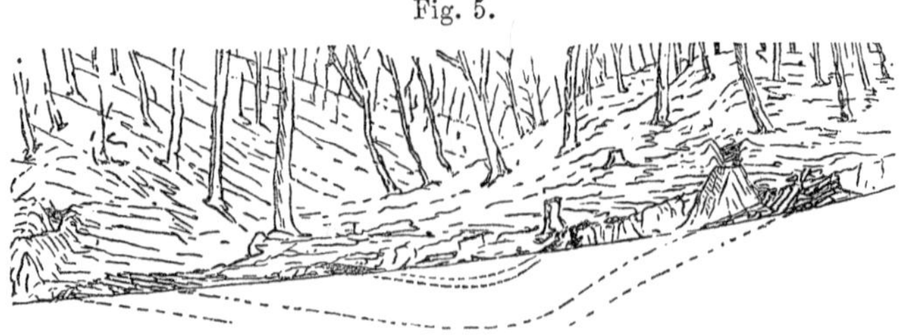

to the coal in place. The whole thickness of the bed thus exposed is seldom visible, part of it being under water, or else part of it covered by the crumbling over of the banks to ascertain the amount of coal. The mistake is constantly committed of sinking a pit in the bed of the stream which the water immediately invades, and the digging is abandoned before the result can be known. A little observation of the rocks in the neighborhood will determine the dip of the coal bed, and a slight exertion of good sense will suffice to dictate the way to open it. If the bed lie flat, or dip up stream, it may be opened a little further down; if down the stream, a little further up, as in the diagram.

17. **Tracing a Bed.**—When coal was first discovered, every hill was supposed to have its own local depots or caverns full. A connection between these separate discoveries was never thought of. To this day, after the growth of a coal trade amounting to nearly a hundred millions of tons a year, innumerable otherwise well-informed and clever people may be found, to whom the notion of tracing the same deposit of coal from hillside to hillside, and across valleys, approaches the absurd. Very few, indeed, properly conceive of deep basins underlying whole counties and carrying down a few unbroken sheets of coal to a depth of thousands of feet. Still fewer can reverse the scene and in imagination restore those gigantic arches which once carried the same beds high through the air from one mountain across to another many miles apart, and are now destroyed and buried up, constituting new sand, gravel and rock deposits in the Atlantic.

And yet in spite of this fundamental error as to the relation of discovered coals, or rather through a natural application of it, another and opposite error has universally obtained. Knowing of no cause for the appearance of coal in one latitude, or on one hillside more than another, the inhabitants of neighboring districts suppose the lines it evidently makes upon the map may be indefinitely protracted, and coal is sought for wherever they thus run. For example, a farmer in Monroe County, Pennsylvania, knows that coal is mined at Tremont, Pottsville, Tamaqua and Mauchchunk, and by the map he sees that a straight line drawn through these four centres, if projected, will pass through or near his farm upon the Delaware. On his farm he finds blackslates, which, although really of a much older formation, he takes of course to be the slates accompanying coal, and persists, perhaps throughout his life, in useless explorations, not knowing that the line of coal, on reaching the Lehigh, turns sharply back and runs toward its starting point. Instances are well known where fortune and reason have sunk together in the search.

But even where the mistake in this its geographical form is corrected it appears again, and much more obstinately, in another topographical form. Coal opened at a certain level upon one hill suggests to almost every mind coal likely to be found *at the same level on the opposite hill*, whereas, in the majority of instances, search for coal based upon this common notion cannot but fail. Of two mountains equally high, parallel, and near together, the one containing coal, people cannot imagine why the other should not contain it likewise, even when, as often happens, they have the plainest evidence before their eyes as they walk or ride along, that the two mountains consist of different systems of rock, one older than the other, and burying itself by the very slant of its layers beneath its fellow. And this is strikingly true of the two mountains which inclose our anthracite and semi-bituminous coal fields in a double frame. The inside mountain alone contains the coal; the outside mountain is the upturned edge of the first of the deep floors of the coal basins.

18. **Dip and Strike.**—Success in tracing coal depends entirely upon one's ability to keep in view the dip and strike of the rocks, and for this end every opening in a bed already found should be made satisfactory on this point; should not be accounted as complete until it exhibited without mistake the *direction and degree of slant* which the bed has, and consequently *the course which a perfectly level gangway in it would take.* The former is the *dip*, the latter the *strike* of the bed. The *dip* of a stratum of rock, slate, coal, or what not, is the line

which a ball would describe if allowed to roll freely down its exposed surface. The *strike* is that line which the edge of a pool of water would leave if dammed up against its exposed surface. The dip and the strike are therefore at right angles to each other, the one vertical, the other horizontal. The drainage, the air shafts, the breasts and coal chutes of a mine follow the dip ; the gangway and galleries follow the strike.

Miners talk of a bed dipping two ways at once, which cannot be. There can be but one gravity line, but one drainage line, but one breast line, and therefore but one dip. Local circumstances confuse a miner's feeling about his bed, and the curves which its irregularities impose upon both the dip and the strike invest it with the same sort of life which a sailor comes involuntarily to respect in his ship. The miner always speaks of his gangway or his " vein" as behaving so and so ; she takes a rise to the one side or the other ; while she pitches as before, she begins also to pitch forward or backward ; and in this way the notion of double dips gets fixed, a notion utterly destructive of all clear geometrical conception of the structure of the interior. A roof dips two ways ; and a hipped roof, while it changes the strength does not the direction of the slant. The sides of a triangular or quadrangular pyramid dip three ways or four. The strata of a volcanic cone dip but one way at any one point, but the dip boxes the compass. It may as well be said of a gun that it shoots two ways at once as of a coal bed that it dips two ways at once. In fact, in the science of gunnery we have our best illustration of the geological phenomena of dip. If a rifle ball be fired from the mouth of a coal mine, while the rifle is held at the angle at which the coal is seen to *dip the most*, and in the same direction, the rifle ball should bury itself in the black dirt of the same bed on the opposite hill. Or if a telescope with a spirit level be sighted from the mouth of a coal mine at a dead level and in the direction of the *strike*, it should enable the rodman to put his target on the outcrop of the same bed on the opposite hill. Such is the theory of tracing beds by strike and dip. Of course in practice the theory is modified by a thousand accidental circumstances which experience and judgment alone can circumvent. Neither dip nor strike is ever perfectly steady and even; on the contrary, they move in infinite variations of the straight line and of the curve, while

Fig. 6.

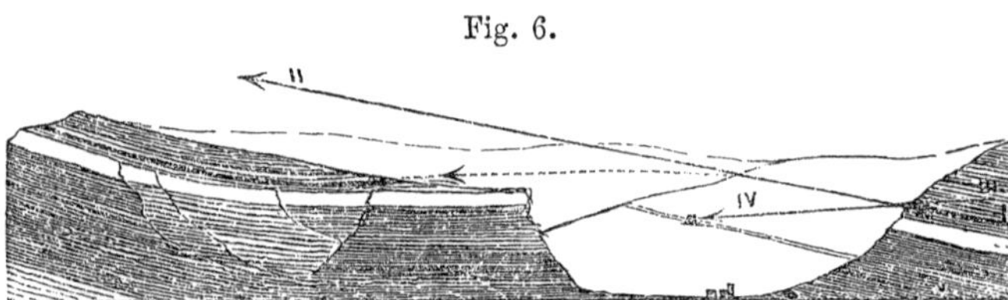

the general course is preserved. The geologist must be always on the watch for eccentricities of every kind. The most unexpected turns and offsets underground are given to the coal and its accompanying rock. It doubles like a hare before the hunter, who is obliged to pursue it by reason or induction, not by sight, and from every unimportant ledge of rocks, from the changing color of the soil, and from the trend of stream and ridge catch indications of its unseen course. (*See* Appendix I.)

19. **Identification.**—It is possible, and in fact it frequently happens with experienced miners, to lose the trace of a bed and pass unconsciously the interval that separates it from another. In fact, it is never certain that two openings may not be upon different beds, until they are identified. And the practical difficulties of identification are so great, that one of the commonest errors of inexperience, prompted and sustained by interest, is that of representing every new coal opening as upon a different bed of coal. It is not too much to affirm that the value of three-fourths of all the geological reports upon the coal regions of the United States, hitherto presented to private owners or joint stock companies, especially those of unsettled parts of the country, is greatly lessened or entirely destroyed by the natural tendency of mind in the geologist to multiply the number of his coal beds. And this disposition is

encouraged by the desire of his employers to receive a good report, by the extreme difficulty in many cases of proving beds to be identical even when the probabilities of it are evident, and by the distance of any future day when actual working or more accurate methods of survey will publish to the world the error. Nor is it a mistake of the vulgar only. The first of dead and living geologists have affixed their signatures to reports in which the evidence of an identity too uncertain to be distinctly stated, but too probable not to do incalculable damage to the extravagant value fixed upon the land, have been therefore wholly suppressed. Lands from which all the beds have been swept off with the exception of the lowest one or two, have been repeatedly represented, by virtue of numerous distant openings on these one or two, as containing as many beds as would suffice to reproduce the whole coal system on the spot. That class of miners who act as pioneers, most of them restless adventurers, related to the settled mining population, as the hunters and lumbermen of the backwoods are related to the settled farmers that succeed them—these men systematically deceive themselves and their employers in this way. Knowing nothing of the general laws of structure, but only of *strike and dip* in relation to the drainage of a single gangway, in a country where the surface is a mass of crags and glens, and covered with fallen rock and impenetrable underwood, they have no means at hand for identifying distant exposures. Every new opening, therefore, which they make is the index not only of a new treasure for the owner of the land, but of a new layer in the system of coal measures.

When a coal has been faithfully opened at one point of its outcrop, its internal constitution and its immediate external relationships must be carefully studied; for in the absence of consecutive openings, or natural exposures, or marked topographical indications, or accurate instrumental surveying, these become the only, and often the sufficient clue to its identification elsewhere. Every coal bed is more or less individualized by some peculiarity of its own. Of two beds ranging widely through a coal region, one will have a characteristic soft broken shaly top, and require more timbering (a fact sure to be an item of practical knowledge with the miners); the other, a firm, unbroken, sand rock roof, self-supporting. One will always show smooth, cultivated, or oak overgrown hill slopes above it; the other pine woods, broken rocky ground, or overhanging cliffs. One will have a large bed of fireclay beneath it everywhere; the other a thin deposit, often almost wanting. One will overlie a bed of limestone, or a bed of iron ore; the other slate or rock. The roof of one bed will be crowded with fossil plants; the other will yield few or none; or one will abound in flattened, or ovate seal-marked, or scale-marked tree-stems, sigillariæ and lepidodendra; the other in ferns; or each will have a different mixture and proportion of these impressions. Sometimes, where there is nothing characteristic in the bed itself, in the fixed number of its plies or layers, or the quality of its slates, or the character of the coal itself (a rare deficiency), there will be a marked character recognizable in some one rock above it or below it; such as the burr-stone iron ore which accompanies the coal bed next below the Lower Freeport coal throughout the valley of the Alleghany River; or the fossiliferous limestone under it; or the two conglomerate sandstone layers which inclose the Mahoning sandstone coal bed; or the pea conglomerate which accompanies the twelfth Shamokin coal; or the cement layers which underlie one of the Bolivar coal beds. These and numerous other instances show that the laws of individuality have governed the deposit of separate coal beds as despotically as the conception and education of men. When all comes to be known, it will appear that our present ignorance alone stands in the way of our recognizing the specific distinctions between the history of each coal bed, without exception, and the rest. Similar epochs returned, but never with entirely identical conditions. In the midst of general resemblances, the course of events was always somewhat different in detail. The state of things which favored the deposit of Clinton Black Slate (lowest part of VIII.) upon the

Oriskany sand-rock (VII.) was rather imitated than repeated in the deposit of the false coal measures (XI.) upon the vespertine white sand (X.), and still later in the deposit of the true coal upon the conglomerate. So the repeated superposition of red shale formations upon the coal of all these epochs was never an exact reproduction of one original, but received some characteristic variation every time it occurred. There must be marks therefore by which we might distinguish every coal bed throughout its whole extent, if we did but know them. In coal regions like those of England and Belgium, where a horizontal system far underground compels the miner to obtain access to his favorite beds by sinking deep shafts through all the strata, and the record of every inch of shafting done for many years has been kept, the identification of the beds by their characteristic peculiarities has reached perfection. But in new countries where the coal is attacked at its outcrop, and tunnels and shafts are little less than the costly expedients of despair, the intermediate rocks are scarcely known, and the cross distance through the strata from bed to bed is unskilfully and inaccurately guessed. This then becomes the ground of topographical science, and the study of the surface by compass and level takes the place of the study of the interior by shafts and cross-cuts; prominent key rocks become of importance, and a geometrical construction of the whole by strike and dip identifies the beds.

But, the same law of individual variation which forbade any coal bed to be exactly like any other above or below it in one locality, was equally hostile to its consistency over great areas. The key, therefore, by which we identify a coal bed throughout a single coal basin gradually fails us, and must be replaced by another, when we attempt its identification at any distance, in other basins, or across great intervals of latitude and longitude. Limestones thin away and disappear as they come east. The remarkable red shale stratum, which everywhere in Western Pennsylvania marks the middle of the coal measures below the Pittsburg coal bed, is scarcely to be recognized on

Broad Top and at Cumberland, and not at all at Pottsville and Wilkesbarre. The Great Mahoning sandstone conglomerate which sealed the close of the lower coal measures, is a great key rock throughout Pennsylvania and Virginia, but is locally modified so as sometimes to be useless. The sand-rock which identifies the two coal beds of Pulaski County, Southern Virginia, over a large area, even where the coal itself varies in quantity and quality, suddenly slips from the observer as he crosses the New River, into Montgomery County, and he must take up another guide. Rules of identification are, therefore, to be applied locally. Each region is to be examined by itself, and its own system of key rocks made out by the comparison of numerous openings, exposures, and rock sections, and the beds to be traced in spite of their transformations by this as an ideal scheme.

But, even after such a scheme is made, the coal-hunter's work does but begin in earnest. His system or "section," made up of alternate pebble-rocks, sand-rocks, clay-rocks, coal slates, and coal, with occasional interpolations of limestone and ore-beds, will present so many *practical resemblances* between its members—many varieties will look so exactly alike at their imperfect exposures, recur so often, be covered up so effectually from close inspection by debris and vegetation, dip in so many unexpected directions and degrees, as to make the discovery and identification of the coals a task to be completed only by the experienced judgment and the patient, accurate, detailed labor of the geologist and his corps. His theory of identification, however correct, will be set at naught by unaccountable and invisible misdemeanors of fact; his scheme of rocks, however carefully completed at one place, will be broken into by additions, or dislocated by omissions at another; his sand-rocks will slide into shales, his conglomerates become fine sands, his favorite coal conceal itself under a degraded type, his never-to-be mistaken limestone disappear, or some new and more important calcareous layer intrude among previously pure clay beds. He must be prepared for any local metamorphosis of every one of his rocks, however astonishing, for any reasonable or unreasonable increase or diminution of his fixed intervals, to say nothing of hidden rolls, increased and even reversed dips, strata thrown over completely on their backs and giving a shemitic direction to the letters on his pages, for downthrows and upthrows, and oblique dislocations of the crust. All these he must accept as new elements in his calculations with the patient submission of a servant, and go over his most satisfactory work again and again, revising, and correcting, and confessing his mistakes, even where he had felt most sure, and perhaps expressed opinions the most confident. He will, indeed, in time, discover certain limits to these annoying interferences, beyond which they are not likely ever to be seen to go, and, as in astronomy, the very perturbations which have so often vitiated his most elaborate results, will in the end serve to lead him through the maze of wider complications, and train him to regard the infinite variety of detail produced by them as the prime beauty and unending pastime of his science. Still, like a magician among his uneasy spirits, he must forever be upon his guard against surprises of all kinds, and expect his embarrassments, conjectures, and discoveries to begin anew at every fresh locality.

In all these respects, however, localities greatly differ. The difficulties of identifying coal are at their maximum in regions of disturbance, in anthracite basins. The upper coals of Pottsville are even in 1856 a mystery. The Sharp Mountain and Mine Hill beds, though separated by but five miles of air-line, have not yet been identified completely. When beds are crushed together, folded up, turned over, and every hillside shows the rocks dipping a different way, the problem becomes of enormous difficulty. In Wales and Belgium, and the many isolated coal fields of the Alps, there is no end to the flexures and abnormal postures of the coal. The Richmond coal basin (of a later day), see Fig. 7, is almost a dead level on the surface, but its beds lying far below the surface are broken in parallel lines and dropped horizontally in successive steps. Such is the length to which this kind of disturbance has gone in the English coal fields, that it is a principal part of the miner's skill to be able to discover, when his gangway suddenly abuts against a wall of rock, whether the coal has gone up or

Fig. 7.

down, whether he must sink his gangway to strike the bed beyond the fault below his feet, or tunnel upwards to find it above his head. On the contrary, the great horizontal outspread of coal in the United States seems strangely free from such throws, and leads us to believe that the land has never been visited with much severer earthquakes than those with which its living inhabitants have grown familiar. It is nevertheless true that our mining operations in these vast fields is hardly yet commenced, and it is a fact that the number of slight downthrows already discovered is considerable.

Every region has its own distinguishing features. It is important, therefore, to find the key to its structure, or rather to its peculiarities; to know what to expect so as not to exaggerate an embarrassment which may be insignificant, nor to be ignorant of hidden natural operations on a great scale which may make, at any moment, the confidence and calculations of the most distinguished stranger ridiculous. So great are these difficulties in the most disturbed parts of the world, that geologists of experience *never* feel at home in them, for they never cease to encounter unlooked for and provoking mechanical anomalies. In such places the whole operation of mining is a perpetual experiment, no one knowing what an hour may bring forth, nor the wisest able to fix a certain value on an acre, a bed, or a gangway. Fortunately these are very local spots; the principal part of the coal in the world is subject to such simple rules of examination and methods of extraction, as fall within the capacity of everybody, and are often spontaneously discovered and applied by men who make no pretensions to science, and whose names have never been heard beyond the miner's hamlet.

CHAPTER II.

THE PLACE OF COAL AMONG THE ROCKS.

1. ONE principal coal era in the geological history of our planet threw into the shade all previous deposits of fossil vegetation, and has never returned. In every age, in every rock formation from the earliest, the impressions of solitary stems or leaves, or bunches of sea-weed, thin flakes of anthracite or exudations of bitumen, the bark of flattened trees turned into coal or charred logs of lignite, tell the same story of a perpetual vegetation covering the earth or growing in the sea under the action of the chemical laws which govern the rise and fall of forests now. In all ages from the beginning carbon has been fossilized as well as lime, iron, phosphorus, or sulphur, fluorine, or silex, in organic forms. In all ages there have been local accumulations of sea-weeds, marsh-ferns and reeds, or plants of a higher grade at the mouths of ancient rivers, in lakes, gulfs, and shallow seas, as now in the Levant, the Mexican Gulf, the Bay of Bengal, the Yellow, Black, and Baltic Seas, the American Lakes, the Gulf of St. Lawrence and the Northern Ocean. While man was absent and the forest grew untamed, wherever great rivers swept over broad continents from lofty mountains, a deluge of uprooted trees and shrubs must have been perpetually sweeping into the sea before their mouths. But as these conditions did not exist until recently, until the tertiary era had almost elapsed—until

the northern half of Asia, three-fourths of South America, one-half of Europe, the margin of North America, perhaps the half of Africa, and nearly all Australia had risen from the waves, when also the forest itself had its floor correspondingly enlarged, we recognize the latest ages of the earth's history as the peculiar time of alluvial lignite or fossil wood deposit. In fact, the present is the age of fossil wood, if ever there was one, especially in the Western Hemisphere. The vast delta of the Mississippi River covering fourteen thousand square miles and ascending its valley almost to the inflow of the Ohio, is known to be made up of alternate layers of sandy, muddy silt, peat bogs, and cypress swamp forests successively submerged. Whether the calculations of Dickinson and Lyell have sufficient data to establish their chronology of seventy thousand years, or not, the lesson taught us by these recent fossil beds of vegetation is the same. It may be impossible to foresee a certain date when the Gulf of Mexico will be contracted to a given limit by the continuance of these alluvia;—its depth, the movements of its volcanically disturbed bed, the deposits of the Amazon and Orinoco swept along into it by the Gulf Stream, the changes that are going on in the land, the clearing of the forest, the increasing violence of the freshets, these are all elements of the problem of difficult acquisition. But no one doubts the fact of lignite deposits on the grandest scale at present taking place between the continent and Cuba, nor of similar and equally immense envelopments of the pine forests of America and Siberia in the silt and shingle of the Mackenzies, the Lena, and the Obi waters of the Northern Sea. And no geologist can forget for a moment that in every geological age the same operations of nature must have been going on *somewhere* upon its surface and upon a scale perpetually and exactly graduated to the ever variable relative proportion of land and sea, the height of mountains, the mean annual temperature and consequent evaporation, the size of water basins, and the force and direction of marine currents.

2. But **Lignite** is not *coal*, nor can the accumulations even

of an Amazon suffice to make a coal bed. Its floating timber may in narrow channels form rafts, like that which obstructed the navigation of the Red River of Louisiana; but when drifted to the sea, each tree must lie alone like the gigantic calamites and firs petrified in the sandstones of the coal measures. The rudest calculation will suffice to set this in its true light. Should the Mississippi send down one tree a minute for a century, with an average length of forty feet, and a foot in diameter, and these be laid together side by side at the bottom of the sea in a single stratum, they would only cover a space of two hundred acres. Were it possible, which it is not, to compress and crystallize these lignites into one stratum six feet thick, they might then constitute a coal bed covering twenty acres. All the forests of the Mississippi valley could not furnish to the sea from their river spoils during a hundred thousand years one of the anthracite coal-beds of Schuylkill County. But if they could, be it remembered that during all that time the river must deposit fifty times the same amount of rock, throughout the whole of which this timber would be pretty equally distributed.

3. On the other hand, the tropical forests of our own day load the actual surface of the earth with what might readily, under proper conditions, make vast beds of coal, although exposure to air, the want of superincumbent pressure, and the infinite ravages of the insect world retard the accumulation, and produce a very different substance in the place of coal. There are, however, places where the needful conditions are perhaps observed; where sub-aerial swamp coast vegetation lays down a submarine soil, while the sea is slowly gaining on the sinking shore, and currents are as slowly keeping up the bottom to its level by the mud and sand which they bring in; places where earthquakes, at long intervals, shake down the crust to a lower level, suddenly deepening the sea, and destroying its vegetation, while in the intervals of rest the shallowness of the waters is restored, and a new vegetation on a higher platform is in time permitted. There are many places in Flo-

rida, Guiana, India, and especially in Middle Africa, where something very analogous to coal must be depositing at the present day. In all ages we must suppose a similar state of things to have locally obtained, but in one era almost universally; this we call the **era of the coal**; it comes midway between the beginning and the end; it is the true noon of geological history, preceded by the Primary dawn and Older Secondary morning, and followed by the afternoon of the New Red, Lias Jurassic, and Tertiary rocks, and by the evening of the Actual times. Nor is this a mere analogy. So universal a phenomenon as the deposit of the coal measures at that epoch seems to have been must have had a well-marked and general cause. We find it intimated by the conditions supposed to be necessary for the formation of coal, viz: a shallow ocean, a tropical swamp vegetation, paroxysmal earthquake subsidences of no great magnitude, and periods of repose. In the more ancient times, the sea was probably too deep and irregular, the shores too steep and unstable. In later times, the pampas and steppes of a levelled ocean had issued too completely into air. The only possible coal era was that which mediated between the early violence and later obstinacy of the ocean bed, tempering the disturbances over vast areas to a regular and moderate repetition, directing shallow circummundane currents far and wide, and lifting the vigor of vegetation to its maximum previous to the gigantic development of animal life which was to follow. It is far from certain that to do this required a special rise of the thermometer, or an unusual evolution of carbonic acid from the interior of the planet into the surrounding atmosphere. Nothing peculiar may have marked the age of coal beyond the mere self-adjustment of the organic action of the crust at this particular point in its series of postures, whereby it offered immense submerged savannahs to the Flora of that day, as it now does extensive aerial plains to the wholly different, but in its way equally prolific vegetable spirit of to-day; the difference being this, that then it could, but now it cannot entomb the results—that is, turn them into coal.

Over extensive tracts of what is now the continent of North America—and the same holds good of the old world—the coal measures were laid down upon a floor prepared for them in two ways, first by a deposit, and second by a denudation; the ocean was filled entirely level, and then afterwards parts of its bed were lifted into the air and again swept off level, and the whole finally slightly submerged. Upon all previous deposits, and upon the upturned edges of some, the coal measures were then laid, to participate, however, in the next commotion; to be themselves elevated into the air and partially swept away. But as the commotions which distracted the surface of the globe while the last coal measures were depositing were of unrivalled, although not unprecedented violence and grandeur, that older ocean so long recipient of older rocks found in the coal its fate, and was drained off to other parts, leaving the coal for the most part permanently under air, and therefore the top and end of the great antemeridian geology. Here and there, indeed, in the lowermost places, the more modern ocean reinvaded it, and deposited tongues or basins of later rocks upon it in estuaries and gulfs; or local subsidings of the crust produced lakes and inland seas upon it with the same result. But so far as the larger and best known coal regions are concerned in this statement, the story of the coal winds up the morning history of the earth, and calls noon to the older workers. One great individual system of rocks is concluded by the coal. A full stop is put in all books here. The earth wears a new look emerging from the tempest that rearranged at this point all the departments of organic life. Everything henceforth is modern. The very first great formation subsequent is called the *New* Red Sandstone. (*See* Appendix II.)

4. The series of which the coal formation is the last has received the now generally accepted name of **Palæozoic**, or the rocks containing the relics of the Ancient Life. Those which follow the coal are called rocks of the Middle Life, Mesozoic. Those which belong to the most recent times and to our own New Life, are called Kainozoic. In the great coal regions of

Fig. 8.

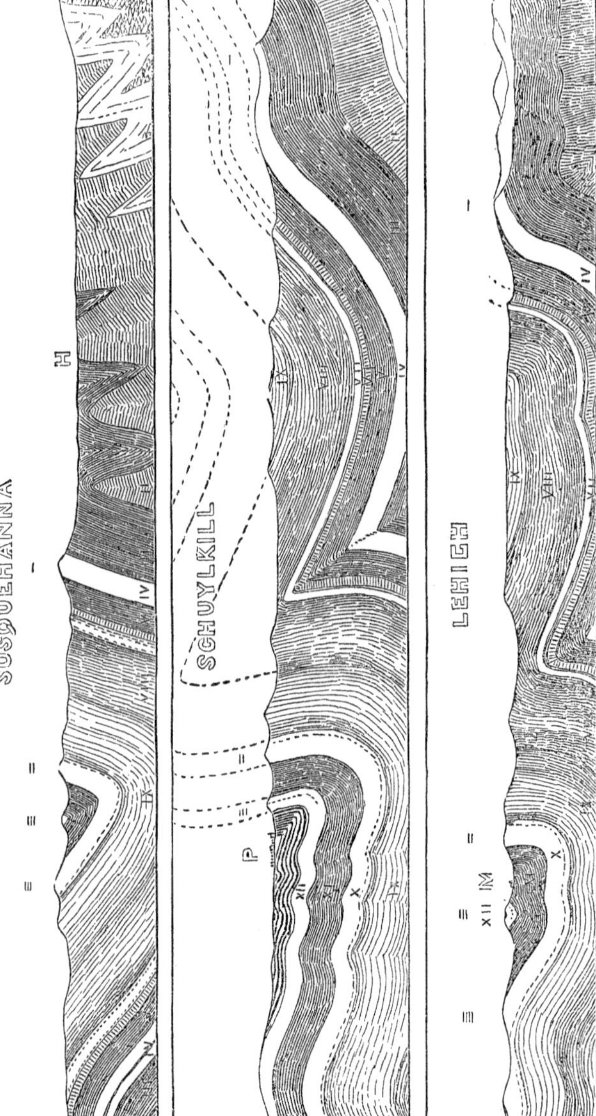

—Blue Mountain. =Second Mountain. ≡Sharp Mountain. ≣Peters' Mountain.
H. Harrisburg. P. Pottsville. M. Mauchchunk.

the United States, we are concerned only with the Palæozoic strata which underlie the coal. They form a floor for it, in some parts seven miles thick. Along the banks of the Schuylkill, Susquehanna, Juniata, Potomac, and New Rivers, they are so thrown up and stand upon their edges that they can be measured directly through from top to bottom, from the coal down to the Potsdam Sandstone, No. I.; and this not once or twice, but many times, as they alternately rise and fall in a succession of vaults and basins; the basins being solidly preserved beneath the surface, but the vaults swept away into the Atlantic. (See Fig. 8, representing three sections north and south through Mauchchunk, Pottsville, and Harrisburg; through the Kittatinny, Second, Sharp, and Peters' Mountains.) The coal measures, where they spanned the vaults, are of course gone. Where they remain, they exist in narrow troughs along the centre lines of the great basins. As we ascend the Atlantic rivers to their heads, and approach the Alleghany Mountain, these vast flexures of the crust suddenly give place to broad low arches, and shallow troughs, permitting the coal measures a wide expanse in each, as seen in the accompanying diagram, Fig. 9, which represents a

Fig. 9.

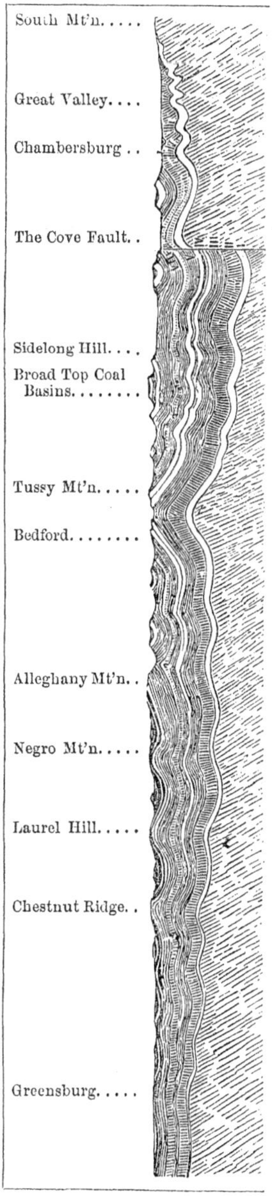

section through Chambersburg and Greensburg, Penn. Finally the whole becomes nearly flat, and the coal spreads out and covers all, far into Ohio, Kentucky, and Tennessee. Then a broad wave brings up the under parts of the floor again across the western parts of Ohio and Indiana, and the coal, of course, is swept away. Beyond this, again, in Illinois and Michigan, and far into Iowa and Missouri, the coal comes on again in fragmentary sheets and isolated patches deposited upon different portions of the floor, according as that had been more or less cut through by the previous action of the denuding forces.

5. This immense floor of Palæozoic rocks consists of four distinct repetitions on a grand scale of conglomerate and sandstone rocks, marking four great epochs of tumult and violence during the Palæozoic age; and of mud and limestone rocks between them which were deposited during long alternate intervals of peace. Each of these **four sand-rocks** marks a new era in the geology, introduces a new creation of plant and animal, and draws its own distinct and separate topographical lines across the face of the earth. That complicated system of mountains which begins at the Hudson and ends in Alabama, obeying to the eye (even as obscurely drawn upon the common maps) evident laws of fixed relationship and parallelism, is seen to be, when studied out, merely the repeated outcrops or edges of these four great sand-rocks as they rise upon the waves or plunge again into the basins side by side. And the valleys which always follow round and forever keep apart the mountains which ever way they run, or however they may double back upon their course or fold themselves in zigzags, are in like manner the outcrops or edges of the softer intermediate limestones, slates, and shales. It is, therefore, impossible to be mistaken in the position of the coal. To reach it or to leave it one must cross a fixed and constant series of well-marked strata, unmistakable in color, composition, grouping, and contents, characterized individually each by its own fossils, metals, and general make, and never out of its rank in the series. To reach or to leave the coal in the centre of the basins one must

pass all the mountain edges of the four sand-rocks, and their intermediate limestone, slate, and shale valleys, in none of which, of course, therefore, can coal, the true coal measures, ever be found. At the present time, these limits of the coal region have become so well and even popularly known, that attempts are seldom made to dig coal in the lower valleys or surrounding mountains. It has become a rule, that "the coal is never seen outside of, or below the Red Shale;" an empirical rule, however, affording little assistance in solving more than one or two of the many questions which arise as to the cause of the presence or absence of the mineral in this or that spot. The limits are, indeed, fixed, but few people can say why. The inhabitants of Sinking Spring or Pine Creek Valley, lying midway between the coal of the Broad Top and the coal of the Alleghany, can see no reason why the summits of the Tussy and Bald Eagle do not furnish an equal supply. The people of Williamsport cannot explain why there should be coal on the mountains back of Dunnstown, and none upon the same mountains, equally high, back of Muncy. The Catskill is higher than the Elk, the Shickshinny, or the Wyoming Mountain, and yet has no coal upon its top. The geologist replies to all such questions, there is an order in the rocks which cannot be discomposed, although it may be and sometimes is interrupted or reversed; and this fixed order assigns to every mountain its contents and even distinguishes the right and left side of the same, its top, its middle, and its foot by a various structure and various constituent elements. Middle Pennsylvania, Maryland, and Virginia and Eastern Tennessee constitute an immense stairway of successive formations, ascending westward from the eastern plains, and from the lowest and earliest deposits to the summit of the Alleghany, to the top of the system, to the highest and the latest formations, to the great western upland, the region of the coal; a stairway which, however warped and dislocated, always shows its original architecture and leads the student from what he knows already to what he is yet to know by an infallible sequence of phenomena perfectly made out and legible by the weakest eye. For the science which at one time was considered the most visionary and immature in its deductions, is now acknowledged to be perfectly sensible and clear in its statements and settled in its practical opinions to the satisfaction of every intelligent student.

If sometimes, as in the case of the anthracite basins or the coals of Middle and Southern Virginia, the coal *appears to be* out of place—appears to lie far in advance of its true position, far east of and in front of the summit of the Alleghany Mountains, where alone we expect to find it first, and therefore apparently down among the lower rocks, a short survey of its surroundings will show how true it is to its place *geologically*, while it has been carried down *geographically* with all the rocks beneath it by extraordinary convulsions of the earth's crust, bent into waves, or parted by long cracks. In every such instance it comes attended with its usual guards, surrounded by the four great sandstone formations, which are seen to stand around it in their order, and with all the other intermediate rocks. These I will now describe.

6. The Potsdam sandstone is the **first great sand-rock** of the Palæozoic system. It spreads into Canada, crosses at the Saut St. Marie, and appears under the coal and limestone rocks of Iowa. It covers the northwest side of the Green Mountains, the highlands, the Easton, Reading and Conewago hills in Pennsylvania, and the Blue Ridge of the South. (No. I.)

Above it are the vast limestone regions of the great Virginia valley, known as the Winchester, Cumberland, Lebanon, or Newburg valley further north—the central avenue and highway of agricultural wealth along the Atlantic mountain border. These are the Trenton and Black River limestones of New York, the limestones of Canada and Wisconsin, and the low limestones on the Mississippi. (No. II.)

Above them lie the Hudson River slates, the slates which form the northwestern half of the great valley just described, and reappear in many parts of the west. (No. III.)

This, which is the *Lower Silurian* system of the English, and made up of the first sand-rock, the first limestone and the first slate of the Palæozoic era, is characterized by peculiar metals and fossils, by lead, specular iron, copper, trilobites, and graptolites, and underlies the United States as a whole; is everywhere the floor; is sometimes, as in Western Pennsylvania and Virginia, three or four miles deep beneath the surface.

There is indeed a **lower system** still, but it is not clearly made out, except in the Lake Superior district, where rocks beneath the Potsdam sandstone contain fossils, and are stratified. Along the western borders of New England, this silurian floor and that lower disputed system maintain a doubtful conflict. Dr. Emmons has identified himself with what all others profess to think absurd, namely, a Taconic system beneath the Potsdam sandstone, and which Rogers, Hall and others describe as lower silurian, *dislocated* and *altered* by ancient volcanic agencies, tossed into awkward postures and robbed of their organic forms. Yet the Rogerses assert that the Potsdam sandstone thickens in Virginia into an enormous congeries of conglomerates and slates, and Emmons appeals to Virginian and Carolinian sections to show a non-conformability. Why the protozoic rocks of Lake Superior *should* not appear upon the Atlantic outcrop beneath the Potsdam sandstone it is not easy to see. But for the present the Taconic system is an Emmonsian myth.

7. The Medina sandstone is the **second great sand-rock** of the four, and the first that shows itself in an independent mountain form. It is a vast precipitate of sand and pebble, the product of some ancient gulf stream or equatorial current, sweeping along the shores of then existing continents. Thin through New York, a mere knife edge of an outcrop along the Mohawk Valley, and scarcely apparent in the west, it begins to thicken as its eastern edge sweeps southward round the Catskill, and rises grandly into the air as the huge Shawangunk Mountain of New Jersey, the Blue Mountain of the Delaware Water Gap, the Kittatinny of Pennsylvania and the North Mountain of Virginia. In all this eastern outcrop,

where it rises steeply out of the ground, its edge is nearly two thousand feet thick, one gigantic plate of stone, and therefore forming a mountain line fifteen hundred feet high, broken at intervals by gaps through which the waters of the inner country break in issuing to the coast. Further south it is thinner, but still forms formidable barriers, like the Peak Mountain, the Clinch, the Walker Mountain in Virginia and Tennessee. How it underlies the Western States, how rapidly it thins away and where its western knife edge lies, whether beneath Ohio and Kentucky or beneath Illinois and Missouri, we can only guess by studying it through central Pennsylvania as it rises like a stricken whale again and again, as if to breathe, before it makes under the name of the Bald Eagle its final plunge beneath the Alleghany Mountains, after which it is no more seen, although it must come near to the surface on the back of the Cincinnati axis, if it has not ceased to exist before it reaches there. All these sand-rocks thin rapidly westward, and not only grow thinner, but finer in their materials, less pebbly and more muddy as if we were getting off from the original shore and far in to the original sea. But this second or "Levant" sandstone, as Rogers calls it, suffers another and characteristic alteration ; it grows double ; two distinct and thinner plates of massive flint proceed from the original one, and being separated by a middle plate of several hundred feet of softer sandy shales and iron ore, form a double crested mountain or a mountain with a terrace on one side ; a feature which distinguishes the mountains of this second sandstone from all others. Lead a geologist blindfold into the Kishicoquillas valley or into Morrison's Cove, and the moment his eyes are opened and his glances fall upon those magical terraces which Taylor and others of the olden school accepted as evidences of former inland lakes, he knows where he is in the order of the rocks, both how near the floor, and how near the coal. One glance at the inverted ship knobs to be described in the following chapter is sufficient.

I have said that Mr. Rogers calls this sandstone formation

"Levant." It follows his Matinal Limestone and Matinal Slates, and marks the sunrise of his Palæozoic day, the opening of the *Upper silurian* era in American geology. But in the earlier nomenclature adopted in the Pennsylvania and Virginian surveys it was called No. IV., and is always so spoken of in common conversation by the members and students of those surveys.

"A mountain of IV." is perhaps the commonest expression in American geology. These mountains are very numerous, being reiterated outcrops or reappearances and disappearances of the Medina sandstone as it rises and sinks in the Appalachian waves. Montour's Ridge, in the forks of the Susquehanna, is the back of such a wave, dying at both ends. The mountains of Union County, the Buffalo, White Deer, Deer Hole, Bald Eagle, Jacks, and Seven Mountains are all mountains of IV. between the Susquehanna and the Juniata. These run on southward into Virginia as Tussey and Brush Mountains. Between them and the eastern outcrops, or the great North Mountain, which keeps on by itself into the south, are many separate canoe-shaped mountains like the Shade, the Black Log, Cove Mountain, and the rest. Those of one group run into each other in curious zigzags, doubling like hares across a thousand streams, always wearing on their backs the red sandstone and red shale, with certain marls, and the rich fossiliferous iron ore from which many of our largest furnaces obtain their stock. (No. V.)

8. In the valleys which encircle these anticlinal mountains of No. IV. and spread out into broad, well watered farming regions, full of hamlets, and traversed by large streams, crop out the rest of the Upper silurian, and the first or lowest of the Devonian rocks—the Clinton, Niagara, and Hamilton groups of the New York geologists, the Meridial and Post-meridial rocks of Rogers' nomenclature, the fifth, sixth, seventh, eighth, and ninth of the original scale, a world of thin limestones, cement layers, black slates, coarse ragged flintstone, many

colored clays, and olive argillaceous sands, thousands of feet in thickness, passing on upwards into red sandy shales and deep red sandstones ushering in our next and third division. It would require a volume to describe these various formations, making up four-fifths of all the valleys of the middle mountain region from New York to Alabama, spreading as limestones over all the surface of the west not covered by the coal, and constituting the west and southern half of New York, the west of Ohio and the larger portions of the other neighboring States. The Niagara river at the Falls plunges over the rocks of the lower part of this great group. Lake Erie and Lake Michigan are merely shallow valleys hollowed out in the middle rocks of this group and permanently submerged. The New York lakes are transverse valleys cut out of No. VIII. It is hardly to be disputed any longer that these are the principal rocks of middle New England, finding their lower level again beyond the Hudson, and descending beneath the coal of Rhode Island and Nova Scotia, but so changed by fire that almost every recognizable feature is obliterated. It is possible that they will hereafter be found to constitute, under a still closer disguise, the hitherto accounted aboriginally ancient rocks beneath the tertiary at Trenton, Philadelphia, Baltimore, and further south, rocks, to which as yet, no age has been assignable. In the interior, however, they are never to be mistaken, either by their relative arrangements, their characteristic aspects, or their wide-spread fossil shells, which are the very same on the Delaware and Schuylkill, as on the Houston and the Clinch. (*See* Appendix III.)

9. It must not be omitted here, for this is its place and it will be referred to hereafter, that there runs through these wide valleys a subordinate range of hills, the out crop more or less distinguished of a peculiar but subordinate sand-rock called the Oriskany Sandstone No. VII. full of fossils, which when once seen can never be mistaken. Its topographical exhibitions are extraordinary. The Pulpit Rocks of the Juniata are frag-

ments of its horizontal layers. (Figs. 10, 11, 12). The rock is remarkable in many ways. It may be said never to vary its character, always a hard rugged cellular iron-stained chert. It is the point or plane of origination for a total change of life, one of the most striking of the phenomena of geology. A new creation begins here. It underlies a black slate deposit, which is the earliest known attempt at making coal.

Fig. 10.

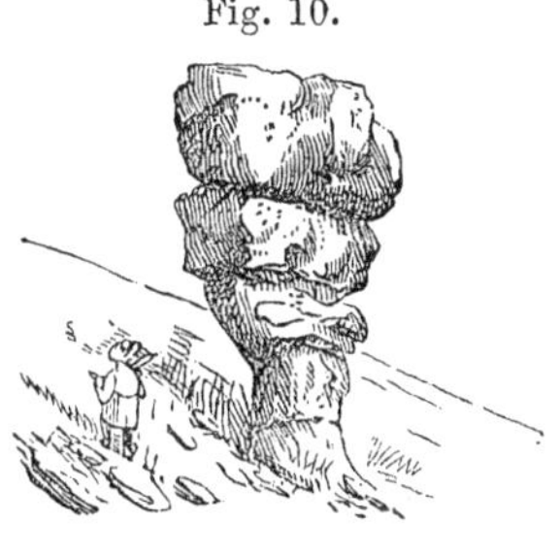

In this last point of view it is of peculiar significance. It will be said shortly how the fourth or highest sand-rock is the base of the coal measures. Here in No. VII. likewise we see a deposit of silex, thin, but of marvellous lateral extent, preceding a thin but equally extensive and consistent deposit of carbon in clay. On the shore hills of Lake Erie this deposit of blackslate actually contains a thin coal bed too thin to work indeed, and interrupted, but nowise distinguishable from the coal beds which long after in the lapse of ages followed it.

Fig. 11.

In the upper valley of the Delaware and Schuylkill near Stroudsburg and Orwigsburg and in many other places, men

Fig. 12.

have spent years, long lives in fact, and fortunes in digging vainly into this blackslate for coal.

10. The **Third Great Sandstone** formation of the floor, No. X., is not so remarkable for introducing a new era of fossil life, as for inaugurating a new system of groups of mountains along its frequent outcroppings, which keep themselves always and everywhere apart from the groups of No. IV., never approaching them within three miles, and usually running in parallel lines with them at a variable distance of from ten to twenty miles, the interval being always filled up with the narrow knobs of the Oriskany, No. VII., and the broad, high, undulating, deeply ravined, and always cultivated hills of No. VIII. On the out or lower side of all these mountains of X. runs an uneven terrace of the red sand of IX.; and in those regions where the rocks stand vertical, this terrace rises to a separate summit, of equal height with the true summit and beautifully parallel with it; a narrow shallow crease divides the double summit, and then the long strait mountain with twin crests of

wonderful evenness, but with this difference, that the outside one is red, and the inside one is white, runs along the map like the double beading of a picture frame. This is true of all that southeastern outcrop which encircles the anthracite coal basins, folding scrupulously in and out around their long sharp points, crossing and recrossing the rivers and creeks, and presenting always outwardly or *from* the coal, its terraces of Old Red Sandstone. The same red and white frame is repeated around the Broad Top Coal Basin south of the Juniata. The Terrace Mountain is its northern point, and western side, and Sidelong Hill its eastern. The same surrounds the Cumberland coal region. This is the formation which constitutes so many of the long straight parallel ridges of Central and Southern Virginia. Like the Medina sandstone last described, it is of immense thickness in the east and thins rapidly towards the west. Its Atlantic outcrop is over two thousand feet thick, hard and white, while its supporting red rocks, No. IX., are at least a mile in thickness; only the upper part of this red mass however forms the terrace or supplementary crest, except where they all lie nearly horizontal. This is the case at the Catskills. Here the vast piles ascend in steps a height of three thousand feet, IX. upon VIII., and X. upon IX., and on the top of all the lower layers of XI. But as we follow this easternmost outcrop south through Virginia into Tennessee it slowly thins as if the original direction of the sediment was from the north and east. Yet more striking is the case when we pass over to its inner outcrops. Around the Broad Top, and where it passes down beneath the Alleghany and the Great Savage it is still a mountain mass, but it rises again in Ohio and Northern Pennsylvania from its underground journey so lean and changed as scarcely to be recognized. It is there a formation of greenish sandstone less than two hundred feet thick. The whole intermediate space, of course, it underlies; that is, all Northern and Western Pennsylvania, all Western Virginia and the whole southern region of the Cumberland Mountain; here it is as thin as in the Catskill region, but here as there helps to pile up the immense plateau, which narrowing as we go southward domineers with its lofty terminal crags the plains of Alabama.

11. Between this Third or Vespertine White Sandstone and the Fourth next to be described, lies but a single formation, the Vespertine red shales of No. XI. This softest of rocks, a pure red mud, once the mere ooze of the gently sloping shore of the quietest of seas, on the ridged wind-wave surfaces of which the rain showers of that ancient day have left their innumerable pits, and the footsteps of unknown lacertian or batrachian tide-haunters are still plainly to be seen—this No. XI. forms a deep valley drained by a succession of short, straight, branchless creeks, which alternately run their red waters opposite ways from high and narrow sheds into the larger streams just where these are cutting through the gaps. One might travel along the string of valleys between these two mountains, as between two of the concentric walls of ancient Pasagarda, round and round the coal basins, without ever seeing a sign of civilization except some clearing where advantage has been taken of the rock riffles in a gap to make a dam and build a saw-mill. There are but four places in Pennsylvania where this sharp straight catalogue of vales belies its nature and widens out into a plain of cultivation; namely, in the Locust valley, behind Tamaqua; in the Catawissa valley, still further north, in Pine Creek valley, west of the Broad Mountain; and in Trough Creek valley, within the Terrace Mountain, south of Huntingdon. Each of them is produced in the same way, that is, by a system of small parallel waves spreading out the rocks into a level floor.

This red shale formation is three thousand feet thick at the Lehigh, Schuylkill, and Susquehanna rivers, and one thousand feet thick at the New River in Southern Virginia. But it rapidly diminishes *across* the measures as we approach the Alleghany Mountains. At Broad Top it is less than one thousand feet thick; at the Alleghany Mountain it is scarcely two hundred, at Blairsville it is thirty feet, and in the Beaver River, lost entirely to view.

Fig. 13.

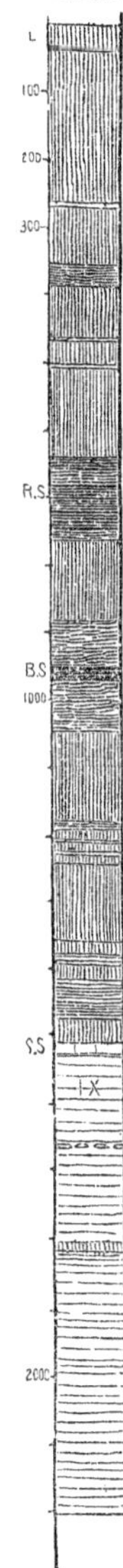

It is impossible with our present knowledge to say positively that the red shale of the accompanying vertical section of the rocks at St. Louis, Fig. 13, struck seven hundred feet down below the level of the Mississippi, is the red shale of XI., but every circumstance is in favor of the supposition, especially the presence of *bituminous shale* two hundred and fifty feet below it, at once recalling the South Virginia coal beds. If so, XI. extends as far as any other member of the Palæozoic system. In Northwestern Pennsylvania a stratum of red shales and reddish argillaceous sandstone occurs everywhere about two hundred feet down from the conglomerate, in the midst of those greenish peculiar looking fossil sandstones which there characterize indifferently X. or VIII. The section at the St. Louis sugar-works has been made twenty-two hundred feet deep with the greatest care, and specimens of every different rock preserved. Those from its middle region are amply sufficient to show the sandstone then bored through to be the Old Red (No. IX.). (Appendix IV.)

No. XI. was peculiarly a shore deposit, rapidly suppressed as it advanced towards the deep sea, into the distant recesses of which only its finest white impalpable particles of fuller's earth was floated, mixed with carbonate of iron. It is, however, not to be regarded as a simple formation for throughout its whole northern outcrop through Upper Pennsylvania, at Towanda, Blossburg, Ralston, Lockhaven, and the Portage Summit, it is a double or triple layer of red shale; and these layers are separated from each other, from fifty to two hundred feet, by greenish sandstones. As it is so important a key or starting-point for an examination of the coal above, this fact must be all the more carefully kept in mind. Two principal events accompanied the deposit of this formation, which closing the Devonian era may be considered as the uppermost third member of the European Old Red sandstone, viz: the embedment of the proto-coal measures, and the precipitation of the Ralston ore; the first event opened, the latter closed its era. Of the latter it is not intended here to treat at large; a chapter might at some proper time be devoted to its extent and character, as developed by the iron works of Lycoming, Fayette, and Westmoreland counties, in Pennsylvania, at Hopewell, Cumberland, and elsewhere, further west.

12. **The False Coal Measures,** as they have been called, the coal of No. XI., the Vespertine coal of Rogers, or as it should properly be called, the PROTO-CARBONIFEROUS Formation, overlies the third great sand-rock of the four, precisely as the blackslate of No. VIII. has been said to overlay the Oriskany sandstone, and as the great Coal measures will be seen to overlie the Conglomerate. This was a second and more successful effort of nature for the preservation of fuel for man, whose coming was foreseen. But still the conditions were not sufficiently fulfilled over the whole area to do more than give promise of a better future. Portions only of the earth were steady enough just at the level of the sea neither to drown the vegetation nor expose its soil. One or two beds, irregular and very thin were everywhere indeed produced and in one region a series of such beds of which two or three are large enough to work. But even these were almost wholly ruined by succeeding earthquake undulations, which slid their floor and roof upon each other, dislocating the layers and grinding the coal to powder.

Everywhere along the inside foot of the mountains of X., from the Catskills to their extreme south limit, and everywhere in the body of the Alleghany Mountain these thin seams have been at different times discovered, and locally noised about. Hunters, lumbermen, and land agents have picked and pried into them. Lands have been sold to Eastern Companies upon a faith in them; but they have never paid. In the gorge of Tipton Creek which descends from the Alleghany

Mountain upon the Little Juniata and the Pennsylvania Railroad near Altoona, one of those beds of coal appears six hundred feet beneath the base of the true coal measures, and nearly three feet thick. Those who believed this to be the lowest of the true coal beds made extensive arrangements for an eastern trade, and justly anticipated a prosperous adventure; whereas the whole carboniferous formation is there not only at the summit, but behind the summit of the mountain.

13. These same beds appear upon the Youghiogheny and Cheat Rivers in Northern Virginia, and at the Augusta Springs in Middle Virginia. But the place to study them is in Montgomery County, further down toward the Tennessee line. As the evidence upon which real coal beds have been assigned a place beneath the great conglomerate has frequently been called for, as well as to exhibit the perfection to which these beds of an untimely birth did nevertheless attain, the following sections, Fig. 14, are given. The lower bed, A, varies

Fig. 14.

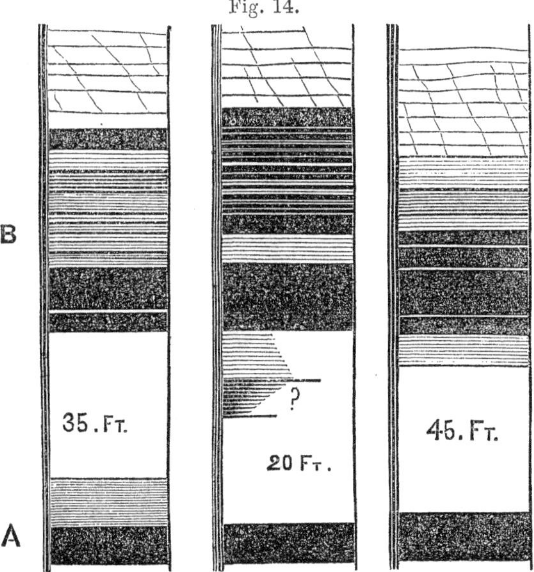

from two to three and a half feet thick and overlies a conglomerate rock about thirty feet. Were there any red shale under *this* conglomerate it would be taken (in the absence of other evidence) for the base of the true coal measures. Even as it is, it requires the strongest assurances of correctness in the observations to induce us to refrain from identifying this bed of coal with the lowest true coal bed which overlies the true conglomerate about the same distance. From twenty to sixty feet higher is the larger bed, B, from six to nine feet thick, composed of alternations of coal and slate, and seldom yielding more than four feet of coal. In this bed again we see a remarkable analogy with the largest second bed of the true coal measures. Nor does the resemblance cease here; for over this upper bed are several smaller ones within a bed of shale six hundred feet in thickness, after which come in a thousand feet of red shale covered with limestone. While in the true coal measures of the North, as will appear hereafter, along the Conemaugh, for example, there are two beds at first, then several smaller ones, followed by the barren measures, from four to seven hundred feet thick, in the upper part of which comes in the red shale bands, which increase in thickness towards Pittsburg, until they occupy at intervals two or three hundred feet of the scale, and on these rest the limestones beneath the great Pittsburg bed. The following diagram, Fig. 15, will show the order of these lower rocks. Beneath the so-

Fig. 15.

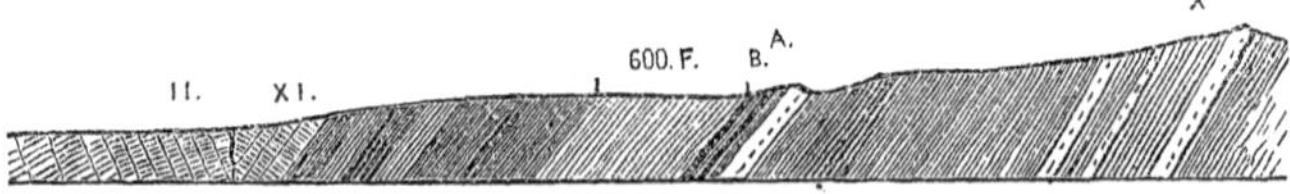

called conglomerate are seen about fifteen hundred feet of sandy shales and massive white sandstones and conglomerates, which are the undoubted equivalents of the Vespertine, Devonian, or Subcarboniferous, No. X. The conglomerate itself, is here a thirty foot plate of coarse sand-rock of very remarkable regularity spreading out evenly over a hundred square miles of surface (except where denuded), but disappearing almost totally after crossing the New River into Montgomery County. Here, in Montgomery County, along the eastern foot of the Brush Mountain back of Christiansburg and Blacktown, the small series of lower coal beds has one of its maxima of value. For thirty miles both beds are workable, yielding a good and tolerable firm coal, lying at low angles (from 15° to 35°) with the horizon, regular, not much broken by faults, and with a breasting from the downthrow on the east to the outcrop on the face of the mountain on the west of from a quarter to a half a mile. Whereas south of the New River (except for the first few miles), all through the Peak Hill country, and along the Brushy Mountain and Little Walker Mountain down through Wythe County on to the Salt and Gypsum mines, a distance of a hundred miles in length, those coal beds are thin, frequently cut off, always faulty and crushed, full of slate, and often rattling like sand when shovelled freshly from the mine.

Fig. 16.

The way in which this coal has been preserved in narrow stripes along the country is shown in the annexed engraving, Fig. 16, which represents a section of the crust, southeast northwest, a little north of Wythe, so as to take in the Peak Hill Basin; but the peculiar topographical features expressed by this interior structure on the outer surface must be reserved for the following chapter.

14. **The Fourth Sand-rock** is the well-known No. XII., or the Great Conglomerate.

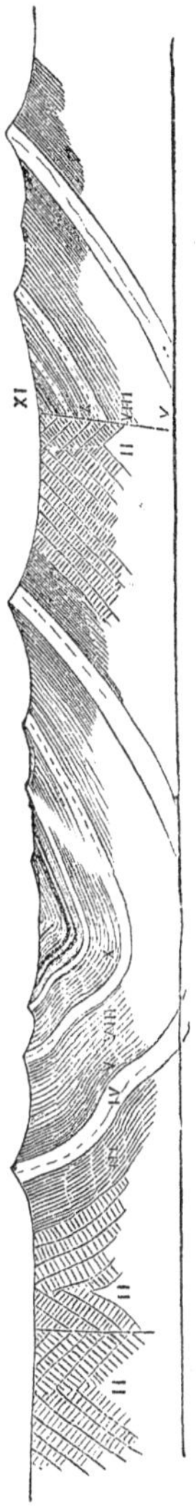

It has its representative in the Millstone Grit beneath the European coal. It is the floor of the true coal measures, an immense preparatory outspread of sand and pebble-stones of every variety, but chiefly pure white quartz; and of every size, from the minute mustard seed and pepper corn to the hen's egg, and in the Susquehanna region even the ostrich egg. A memorable attempt has lately been made by men in the West, not to be despised, nor their arguments overlooked, to prove these pebbles *concretions*. The suggestion, however startling, has probably been made to every mind at some stage or other of its inquiries into the geology of coal, but has almost always been rejected at the first return of reflection. The evidence of the rolled origin of these pebbles is overwhelming; from their shape, that of river cobble and brook stones, and of the constituent parts of ocean current banks and diluvial terraces; from their constitution, without nuclei, diversified in elements and color, and bearing none of the marks of segregation; from their local relation to one another, and to the tree stems and fuci packed up with them in the block, evidently a heterogeneous mass; from their law of distribution, which reproduces all the facts we are familiar with in oceanic shingle beds, thickening in one direction and thinning in another, the local size of the pebbles agreeing with the local size of the bed, and therefore with the local turbulence of the current which held them in mechanical suspension. Everything in fact assures us against the notion of the concretionary origin of these pebbles. Nor is there any sufficient reason for selecting the pure white quartz pebbles from the rest, as segregations, for in all other respects they resemble the rest as rolled stones.

There are nevertheless certain certified facts which vindicate this notion of concretion from the charge of absurdity, and make it of importance as a hint to examination in some new direction, we know not what as yet. The principal one of these facts is one that every geologist must verify for himself, and it opens, when so verified, a world of conjecture. The author has repeatedly seen in the field, and more than once in the presence of a geologist of deserved celebrity, M. Desor, the impression of a plant upon the flattened surface of a pebble. He has therefore no hesitation in republishing this figure of some quartz pebbles impressed by plants, from a pamphlet published at

Fig. 17.

Cleveland by Prof. Jehu Brainard, embodying a paper on the subject presented by him with specimens at the Cleveland meeting in 1854, in which Mr. Brainard argues strongly that the quartz conglomerates of the coal measures must be excepted from other and common conglomerates; denies their heterogeneous constitution; denies the admixture of primary pebbles, trap, homestone, or other rock; insists upon the immense outspread of thin seams of pebble between seams of clay without alternations or gradations; considers that successive layers of lime and quartz rock are separate chemical precipitations from an ocean furnished with decompositions of feldspar and mica; contends that all crystals must show their angles at even the last stages of detrition; refers to the absence of a cement to show that the silex was in solution; accounts for the pitted surface of the pebbles, which he thinks an argument against water polishing, by a surrounding zone of sharp crystal ends; exhibits pebbles with well formed crystals inside of them, and in one case a crystal projecting from [sticking to ?] the side, which therefore he argues could not possibly have suffered detrition; pebbles that have the impression of plants upon them; pebbles in the hollow body of calamites and other fossil trees; and winds up with the following sentence:—

"I have traced this bed [a thin band of white quartz pebbles lying immediately upon a soft, dark, deep-water, mud deposit] for more than one hundred miles and find this feature uniform. But what is still more unaccountable upon the old hypothesis, is the fact that the pebbles are flattened upon their under surface, exhibiting unequivocal marks of their having been deposited upon the smooth surface of the shale in a soft and gelatinous state, where they remained in a perfectly quiet state until induration had taken place, and from which position they have not since been disturbed, as seen in the annexed figure."

Fig. 18.

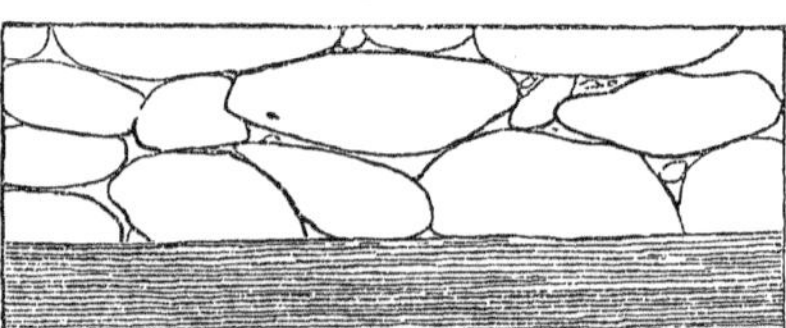

It is always needful to meet the calm statement of a new and apparently absurd theory in the same spirit in which it is offered. But the above arguments answer two very different, and both of them important ends.

In the first place they show how dangerous it is to argue from *local* phenomena to universal conclusions. The heterogeneous character of pebble rock may be denied in Ohio, where once the deep ocean was, into which only the small round pea-quartz of that storm of mingled mud and stone could reach, long after all its accompanying clay-stones had been triturated and dissolved; but how impossible it is to deny it, in Schuylkill County, Pennsylvania, where the pebbles of hard red sand are to be seen side by side with those of milky quartz, and rolled into depressions, scoured out of the upper surface of the rock next beneath the one they form. The same current which carried forward these minute pebbles planed off also the clay on which they were to rest; and it has long been a pleasant subject to observe how the pebbles mostly lie on their flat sides like oyster-shells. The Cape May diamonds found on every shore do not retain a trace of original crystallization and have their surfaces thoroughly ground and pitted. In Ohio the cement may be and is slight; it is less everywhere in this formation than in most other conglomerates; but it is not wholly absent anywhere, and is oftentimes abundant; its comparative absence is the cause of the characteristic whiteness in the cliffs which this formation forms, and in the sand which as if in mockery of the shore, or in some reminiscence of its own origin, it spreads over the tops of all its mountains. Crystals occur in cavities in large masses of rock, and of course some of these masses when broken and rolled down to the dimensions of pebbles would preserve a cavity or two with the crystals intact. As it is well known that the stems of trees growing in coal beds decayed within, while their bark was being buried up in pebbles, sand, and mud, which when it reached the top of the broken trunk poured in and filled the interior, it would be strange if pebbles were not sometimes found within their silicified casts.

But after disposing of all these arguments, *concretionary* quartz remains a possibility, and some as yet unknown method of explaining its susceptibility to *impressions*, is a great desideratum. Prof. Brainard acutely refers to the soft form of silex in the laboratory, making both concretion and impression possible. It is better, perhaps, to refer to the known fluidity of all matter, glass, lens-metal, and the hardest substances under sufficient long continued pressure, and suppose some infinitely slow impression made by plants and by each other upon the stones.

It is characteristic of certain beds of coal conglomerate, and that over wide areas, that they offer to the eye a certain *confluent* structure, while others are equally characterized by the universally distinct isolation of their pebbles, each one accented, as it were, while the other variety badly pronounces its constituent parts. This is very distinctly to be observed, for instance, in the Broad Top coal region, the lower coal-conglomerate of which is confluent, while the upper or Cook rock, both here and along the Alleghany Mountain, is finely discriminate.

7

15. It is certain that the origin of the conglomerate was oriental. It must have been produced along the shores of land which at the date of its deposit bounded the ocean in which it was deposited upon the east. This ancient ocean covered the United States; the shore perhaps ran somewhere along from

Fig. 19.

New York past Richmond southward; until we know more about the Baltimore and Philadelphia rocks, and on account of the denudation which has taken place, it is impossible to say exactly where. How much the Philadelphia or Baltimore hills were reduced in size by that subsequent denudation is unknown; nor whether the Atlantic was then continental or was already then, or always had been sea—as Dana thinks. This much only is certain, that the conglomerate is of great thickness and very coarse towards the southeast, and grows thinner, finer, and more purely siliceous the further it is traced northwestward. In the accompanying woodcut (Fig. 19) it is seen consisting principally of one horizontal plate, fragments of which tessellate at intervals the summit of the Towanda Mountain in Northern Pennsylvania. This plate is fifteen feet thick, occasionally subdivided into two, and always under and overlaid by thinner plates, the whole measuring less than a hundred feet. At Pottsville, on the contrary, the formation measures fourteen hundred feet. Upon the Alleghany waters, and the branches of the Beaver River, it consists of two principal plates of massive square-grained sandstone, sometimes studded with small white quartz pebbles, and sometimes weathered full of holes, or into a curiously ridged surface, the ridges forming a massive reticulation in bas-relief. The most remarkable specimens I have ever seen of this surface-weathering were on the branches of the Kenawha in Western Virginia, one of which, a mass weighing several hundred pounds, was presented by Mr. Edward Mitchell to the Natural History Society of Philadelphia. There also the whole formation is about a hundred feet thick. On Broad Top, two hundred miles further east, it varies from one to two hundred feet. We should therefore say, at first sight, that the increase westward was a local fact confined to the anthracite and semi-anthracite region. But in fact, what we call the Conglomerate Proper in that region is but one, and the lowermost, of several masses of sand and gravel thrown down at intervals upon each other, with coals and clays between; whereas the conglomerate of the west includes, perhaps, several of these masses, or what is left of them, each one having thinned out in that direction. At Pottsville, and through to Shamokin and Hazleton, there are four massive conglomerates above the Conglomerate Proper, each one from forty to eighty feet thick, and several massive sand-rocks besides of equal thickness, over these again. In Broad Top and the Cumberland region, these upper conglomerates and sand-rocks of "the coal measures" are also present, but all of them thin. Further west there is a certain declension, but its regularity or irregularity we have as yet no means of determining; that will be left as a legacy of research for the explorers of a future generation. This much, however, is very certain, and should excite our admiration as one of those curious coincidences which may well bear the name of providence, and be received as evidences of the forethought of benevolence, that we are indebted to this enormous local eastward thickening of the Conglomerate Proper and the conglomerate and sandstone beds above it, for our anthracite treasures. Had the rocks beneath the anthracite coal been the mere thin sheets of sand which they are to the westward, weakened still further by intercalations of clay and coal, their outcrop edges never could have withstood the rush of denuding waters, and protected as they did the mineral fuel within their gigantic folds. What now are groups of long, slender, united, or closely parallel coal basins, would have been, but for this protection, wastes of red sandstone, or deep lakes in the olive shales of No. VIII., like those of the north. The comparatively little coal that has been hardly left in these small basins would then have gone the way of all that vast original deposit, the debris of which lies buried under the profoundest bottoms of the Atlantic, together with the immensely greater ruin of the formations underlying and preceding it. This will perhaps be made plain by the following map (Fig. 20), which will also be our best introduction to the coal formation.

16. **The Coal Measures.**—The map is intended to exhibit

Fig. 20.

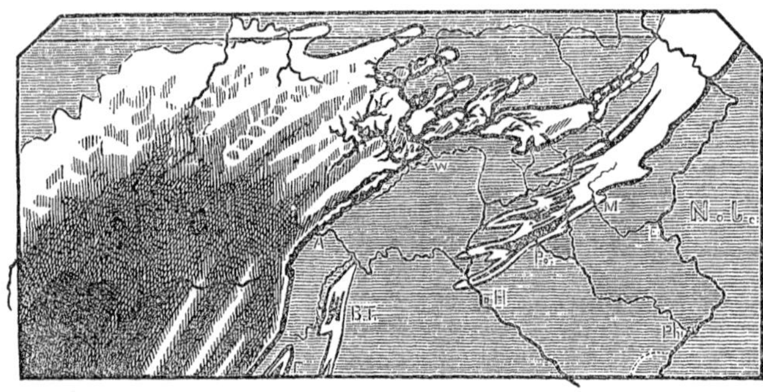

the northeastern portion of the great eastern coal field of the United States and the agent of its partial preservation, to wit, that great plate of formations lying beneath it, by which it was to a certain extent protected, and upon the fragments of which its own patches lie in basin form. The part ruled a half tint represents the formations from I. to VIII. from the Potsdam sandstone up to the Chemung group. The ragged portion left white, and over which the darker tints are spread, represents the combined thickness of IX., X., XI. and XII.; that is, of the upper two great sand-rocks of the four described in the foregoing pages (IX., X.) and (XII.), together with their included shale (XI.), and the coal above. The letters F, B. T, A, W, H, Po, M, Ph, N. J. stand for Frankstown, Broad Top, Altoona, Williamsport, Harrisburg, Pottsville, Mauch Chunk, Philadelphia and New Jersey. The band of white running in at the northeast corner and crossed by the Delaware River is the Catskill Mountain plateau; on which as it declines southwestward the anthracite basins are seen stretched out lengthwise within its close, sharp folds. As the Elk Mountain, it crosses the North Branch of the Susquehanna; and then as the Mahoopeny or North Mountain, it curves into the great Alleghany Upland, and so continues, widening and deepening on into Virginia. The Frostburg basin is attached to it, but the attachment is below the map. The Broad Top basin is the only fragment containing coal wholly separated from it, the back of a high anticlinal wave elevating itself between them. Three other similar and parallel but lower anticlinals further west are also seen running up through it denuded of their coal. These pass on side by side into Virginia. Of these the first one Negro Mountain soon dies down beneath the coal or is continued past the viaduct. The other two Laurel Hill and Chestnut Ridge run on far north of Johnstown and Blairsville and until they are seen issuing as broad anticlinal valleys, between high synclinal mountains upon the rolling plain of New York. Two other anticlinals lower than these sweep round outside of them to the west before we reach Pittsburg, and issue at similar valleys in Potter and Bradford Counties between other similar synclinal mountains, of which therefore there are just five, the titanic fingers of the great hand of coal expanded towards the north. Each is the terminus of a long trough-shaped plate of IX., X., XI. and XII., about ten miles wide, curving gently round into Virginia, descending (relatively to tide level) by insensible gradations as it goes and protecting within its limits more and more of the coal upon its top. At first at its extreme end it retains only a single patch of the lowest bed a few acres in extent; then two or three; then larger patches of the two lower beds with wide intervals of barren ground. Gradually it permits these fragments first to join each other and then to pass over the anticlinals bounding it to coalesce with those of its neighbors. It is soon seen holding the whole Lower system of beds; and now a broad waving sheet of coal measures spread across all the basins. These go on declining—and receive upon this first system of rocks a second called the Barren measures, because destitute of workable beds, many hundred feet in thickness. By the time they cross the Pennsylvania Railroad they begin to receive (still first along their centre lines and afterwards throughout their length and breadth) a third division of the carboniferous formation the Upper coal system, beginning with the Pittsburg bed. This is represented nearly black upon the map. In the extreme southwest corner of the State a fourth subdivision or second set of barren sands and shales come in to cover all.

Fig. 21 represents a section across the northern finger mountains, or terminal basins, in Schuylkill, Luzerne, Lycoming, Bradford, and Potter counties, Northern Pennsylvania. (Fig. 9, on page, 53, shows the same basins, further south, and filled with coal.) The coal is seen on the tops of the mountains—first in the Mauch Chunk, or Pottsville Basin (on the left), next on the Beaver Meadows, Hazleton and Black Creek basins, and others of that group; then in the little outlier of McCauley's Mountain; then in the Wilkesbarre basin; then in the high, shallow, terminal, bituminous basins of which we have been speaking—the Mahoopeny, the Towanda, the Blossburg, the Crooked Creek or Wellsborough, and the Cowanesque.

The extreme southern portion of the great coal field exhibits a similar structure; but the rise of the basins is there towards the south, and the long, synclinal finger mountains, point towards the Gulf of Mexico. The Nashville and Chatanooga Railroad crosses them at their narrowest place, but their final termination overlooks the country south of Huntsville, in Alabama. Local differences of magnitude present themselves between the opposite ends of this great field; the concave structure is more decided at the north, and a disposition to flatness and upthrow is characteristic of the south; the north finger mountains, and included valleys, are all sand and mud, while limestones form not only the foot but more than half the flank of those towards the south. But the same principles of action ruled in the preservation of the coal system, so far as it has been preserved, from the Great Bend to the Muscle Shoals. These principles, the laws of denudation, will be the subject of another chapter. In this one we have to do only with the order and posture of the coal.

Fig. 21.

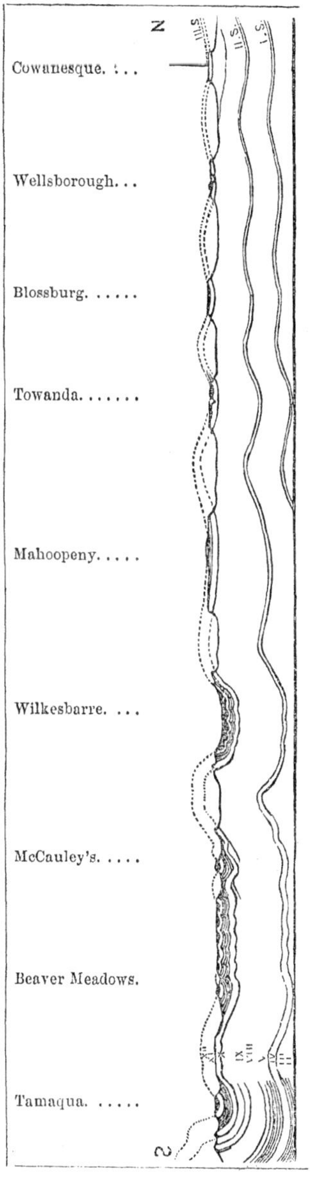

Cowanesque. . . .
Wellsborough. . .
Blossburg. . . . . .
Towanda. . . . . . .
Mahoopeny. . . . .
Wilkesbarre. . . .
McCauley's. . . . .
Beaver Meadows.
Tamaqua. . . . . .

17. In England, as in this country, the coal measures are piled upon Millstone Grit. The American traveller who climbs one side of the gorge in which are the coal mines of Allier, in the South of France, and looks across upon the nearly vertical ribs of conglomerate which mount the other before him, can with difficulty persuade himself that some incantation has not transported him back across the sea to Mount Carbon, and seated him above the old gangway of the Baracleugh. The Schuylkill rushes below him through the gap; before him rises a twin to the Sharp Mountain; he recognizes the same rocks, the same dip, the same conglomerate, the same entries, the same trains of cars, running in the same direction. The only difference which he can discern, does but confirm the resemblance. Here it is to the left instead of to the right, that is, above the coal instead of beneath the conglomerate, that the roads and fields are red. But if acquainted with the Virginian coal, he only argues from this fact, that the beds upon which he looks belong to the proto-carboniferous, rather than to the carboniferous era.

In Chili, and it seems in all parts of the earth, any great deposit of coal began with tumultuary movements of the earth and sea, the only explanation of which, now feasible, appears to come from the known reaction of strained matter, by which, when fracture is once accomplished, unstable gives place to stable equilibrium of all the parts, and there follows, consequent upon short passionate accesses of earthquake commotion, long periods of repose. It is the best explanation we have; but yet not wholly satisfactory. For we see that the early coals, which were the largest, were formed in the very midst of and between the monuments of tumult, the conglomerates themselves; and that these attain their maximum of development precisely in the most disturbed locality, that of the anthracite; while, on the other hand, as we ascend the series and see the softer rocks multiplying, and at last occupying the whole ground, we see them barren also of the coal. These softer rocks, which certainly are as much monuments of a tranquil time as conglomerates are of a tumultuous state of things, contain but thin impoverished layers of blackslate. Just when we should expect the beds of coal to be the largest and most numerous they vanish from the scale. Not until we have gone up through the barren measures and begin to enter what may be called the third or limestone era do we find again deposits commensurate with the profound tranquillity which seems at that time to have inspired nature in this western world, a tranquillity as treacherous as it was profound, for it lasted only long enough to prepare the elements of the most thorough change which our geology has ever undergone, the plication of the crust, the emergence of the continent, the destruction of, perhaps, a moiety of all the coal originally deposited, and the final shaping of the present surface. It may well be doubted, therefore, whether in view of all things we have yet obtained the right clue to the genesis of coal, or more than a distant bird's-eye view of the circumstances under which that remarkable era was ushered in and carried through.

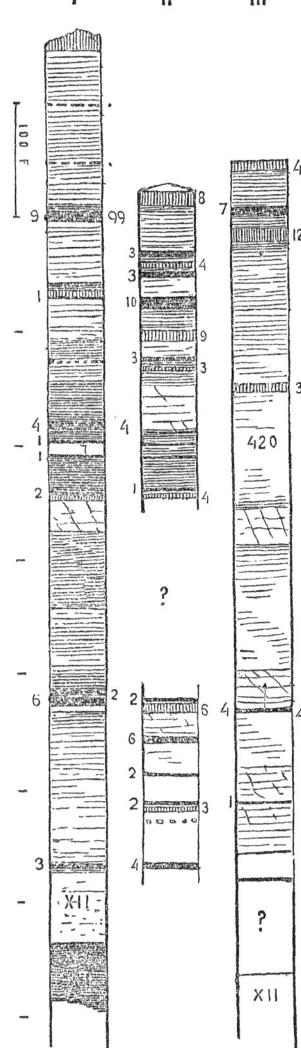

18. **The Bituminous Coals.**—The best illustration of what has just been said will be found in a series of vertical sections (VIII.—XIV., XV.—XXII.) of the coal measures of Fayette and Westmoreland Counties, made fifteen years ago by Dr. Robert S. M. Jackson, of the Pennsylvania Survey, and not yet superseded by anything better on that ground. The observations were made along the western flank of Chestnut Ridge from Blairsville to the Maryland line, and the sections are here given to exhibit the *triple* or rather quadruple subdivision of the carboniferous system. Their other features are of local interest, but not essential to our subject at this point. (*See* Appendix V.)

To show, however, how these local traits do vary at short distances, other sections (I. to VIII.) are given from neighboring counties (Somerset and Cambria). From all these sections seen at once the eye gathers at a glance the prominent facts of prominent interest.

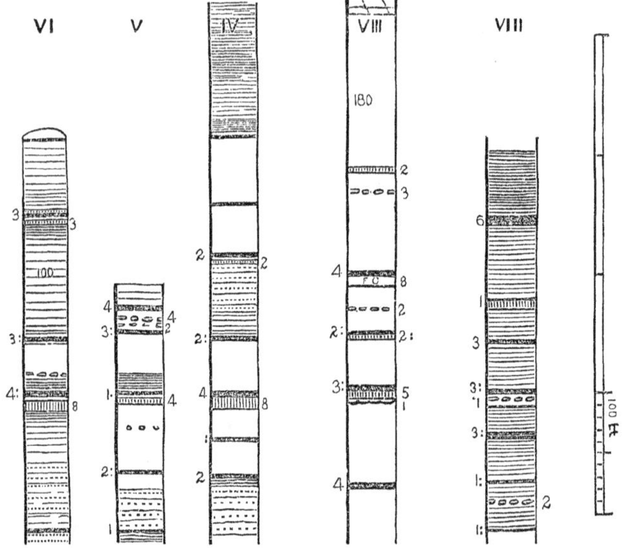

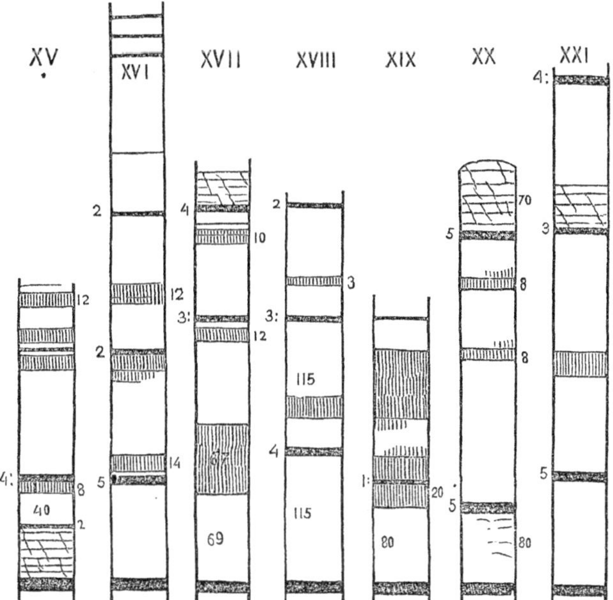

Sec. I. Elk Lick, Somerset Co., between Negro and Alleghany Mountains, Pa.

" II. Ligonier, Fayette Co., lower part at Lockport, &c., on Penn. R. R.

" III. Blairsville, Indiana Co., Penn. R. R.

" IV. (VI.) Indian Creek, Somerset Co.

" V. Laurel Run, Somerset Co.

" VI. (IV.) Castleman's River, Somerset Co.

" VII. (VIII.) Conemaugh River, Cambria Co.

" VIII. Clearfield Co., Pa., Susquehanna, West Branch. (J. L. Hodge.)

" VIII. (VII.) Virginia Line west of Chestnut Ridge. (R. M. Jackson.)

" IX. Redstone Creek.

" X. Jacob's Creek, pages 54, 130, fourth report.

" XI. Jacob's Creek.

" XII. Jacob's Creek.

" XIII. Big Sewickly.

Sec. XIV. Loyalhanna.
" XV. Above Union, Fayette Co., Pa.
" XVI. Redstone Creek.
" XVII. Redstone east of the axis, p. 180, fourth report.
" XVIII. Redstone west of the axis, p. 179, fourth report.
" XIX. Jacob's Creek, p. 130, fourth report.
" XX. ————, p. 116, fourth report.
" XXI. Big Sewickly Creek.

One coal bed larger than the rest runs through all these sections at an elevation of six or seven hundred feet above the conglomerate. This is the **Pittsburg bed,** a deposit of such extent that we have identified it from Cumberland in the east to Wheeling on the Ohio, a breadth of a hundred and fifty miles, and at points equally distant from each other north and south. If it can be identified in the anthracite basins also, which we have everything but an absolute demonstration for believing, its outspread will be doubled. Taking into account the immense area over which it must originally have stretched to include these scattered points, an area little short of two hundred thousand square miles, it presents itself as one of the most remarkable and significant facts of science, a starting point for many theoretical speculations, and, practically, a sure horizon of observation from which to measure and identify the rocks above and below it. It has been made, in fact, the base line of our carboniferous geology. Its size, constitution, quality, and accompaniments are all unmistakable. Its size throughout the major part of its present extent is eight feet, gradually increasing eastward in Ligonier Valley (Fayette) to nine, in Elk Lick Township (Somerset) to eleven, and in the Cumberland region to fourteen feet. Its constitution is everywhere that of a double bed, two or three feet of bony coal above, and five to ten feet of rich, solid, pure bituminous coal below, yields forty to forty-five cubic feet of gas to ten pounds of coal, where it is best known, and is very free from sulphur; although strangely enough both its roof and floor are full of pyrites. As to its accompaniments, they may be seen at a glance upon the sections; above

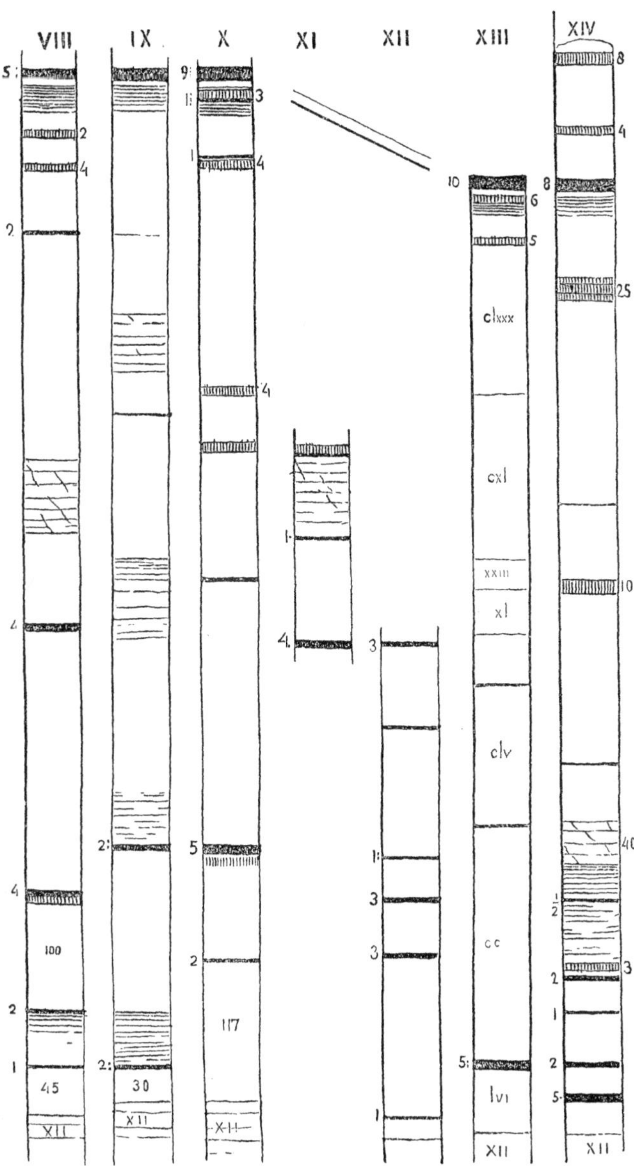
VIII
IX
X
XI
XII
XIII
XIV
clxxx
cxl
xxiii
xl
clv
cc
lvi
100
45
30
117
XII

it are massive limestones and the workable coal beds of the Upper Series, of which it forms the base; and below it are the Barren Measures which separate it from the Lower Series of coal beds resting upon the conglomerate.

19. **The Anthracite Coals.**—Similar sections* made in the gaps of the Sharp Mountain, in Dauphin and Schuylkill counties, by Pinegrove and Pottsville, compared with others at Shamokin, Wilkesbarre, Pittston, and Scranton (XXII.—XXVIII.), will not show this triple structure so plainly. The coal beds have, in this direction, greatly increased in size, and apparently in number, by which we may perhaps understand that the numerous seams of black slate scattered through the whole carboniferous system in the west, too thin to be exposed except by accident, of too little value to be sought for, and too insignificant not to be overlooked in any but a careful enumeration for theoretical purposes—find at the east their maximum of size and purity, and thereby interfere with the identification of the western workable beds of coal, among which, in the imperfect sections of the western measures, these slates do not appear. It has been energetically stated as an impossibility to identify the anthracite coal-beds over distances even as short as that from Soloman's Gap to Scranton, from Pottsville to

* Sec. XXII. St. Clair near Pottsville, Schuylkill Co., Pa. (P. W. Shaffer).
" XXIII. Lee's mines, below Wilkesbarre. (A. McKinley.)
" XXIV. Wilkesbarre section. (Logan, McKinley.)
" XXV. West Pittston boring; Wilkesbarre basin.
" XXVI. Griffin Lot, Scranton, W. basin. (Needham.)
" XXVII. Diamond Mine survey, W. basin, near Scranton. (Needham.)
" XXVIII. Scranton, Wilkesbarre basin. (H. D. Rogers.)
" XXIX. Shamokin. (H. D. Rogers and P. W. Shaffer.)
" XXX. Gold Mine Gap, Dauphin Co., Pa. (R. C. Taylor.)
" XXXI. Raush Creek Gap, " " in the Sharp Mountain. (R. C. Taylor.)
" XXXII. Mt. Eagle and Black Spring Gap, in the Sharp Mountain. (R. C. Taylor.)

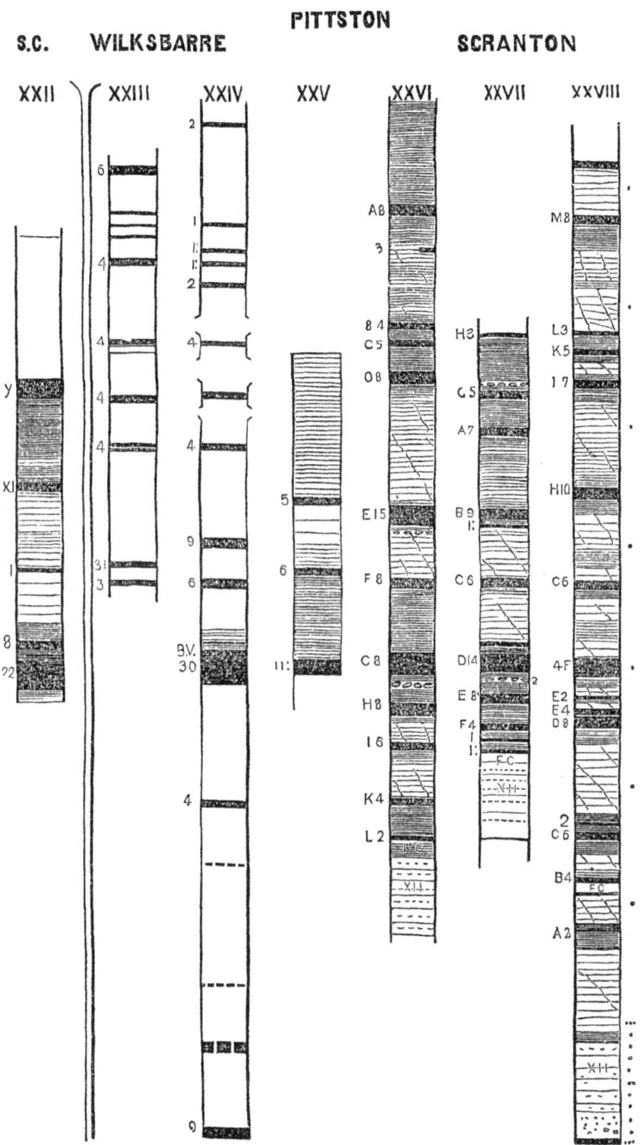
PITTSTON
S.C.
WILKSBARRE
SCRANTON
XXII
XXIII
XXIV
XXV
XXVI
XXVII
XXVIII

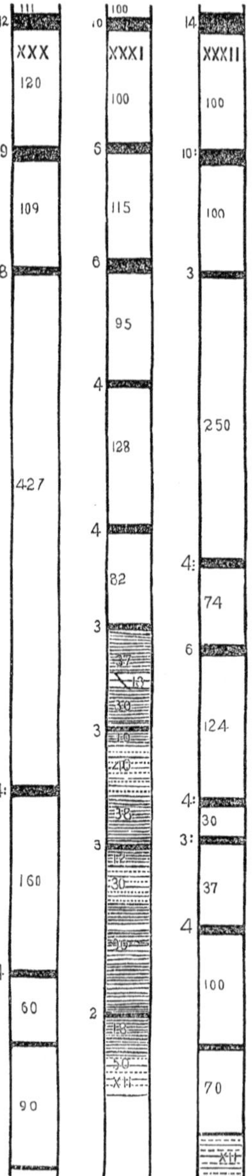

Tamaqua, or from Beaver Meadows to Wilkesbarre; but the juxtaposition of a few carefully prepared vertical sections is sufficient to show how easily in fact it may be done. Such sections are the growth of time, the fruit of repeated and expensive surveys of the same ground, constructions piecemeal by successive observers, one correcting and complimenting the data of another. But ten years ago, when the author was computing the suite of maps and sections which were to illustrate Prof. Rogers's final report of the survey of Pennsylvania, imperfect as the sections constructed at a still earlier day by a great variety of hands and some of them entirely untrained necessarily at that time were, he saw all the radical elements present in them for an identification of not only the anthracite beds among themselves, but of the anthracite with the bituminous beds of the far west. Since then, more carefully constructed sections have been obtained by numerous private surveys, and there can be no reasonable doubt that the time is not far distant when every persistent coal will be known wherever it outcrops, from Carbondale, Meadville, and Massillon, to the southern limits of Tennessee. And if so, it is not too much to expect the same success with the scattered fields of Illinois. The coals of St. Louis and St. Clair are evidently the beds of Iowa. Even some of the innumerable layers of Nova Scotia coal may, in time, be found to be the same with some which crowd the Pottsville basin. Here, however, we must stop—on the extremest verge of possible identities. To speak of the English "High Main" and "Low Main," in Schuylkill County or at Broad Top, is absurd. Even for what is possible we must await the results of the labors of Leo Lesquerox, of Columbus, Newbury, of Cleveland, Dr. King, of Greensburg, and others who are devoting themselves to the botany of coal.

20. **Fossil Plants** must be our guide at last. Wherever across an interval of denudation, such local changes occur in the character of a bed, or of its accompanying rocks, as to make identification impossible, we may still have recourse with confidence to the species and groupings of its plants. Where continents or oceans intervene we have no other materials even to build conjecture of. Unfortunately as yet almost nothing has been published. We wait for volumes to appear which have been years compiling, and are already written, and for crowded cabinets yet to be assorted. Newbury has collected over two hundred American species of coal plants. Lesquerox has described (in the vi. vol. of the *Boston Journal of Nat. Hist.*, p. 409), a hundred and ten out of more than two hundred American species which he has laboriously studied on the ground, in a systematic examination of all the separate beds throughout the coal fields of the Eastern and Middle States. The great Brogniart, because his specimens came from all quarters of the globe unlabelled for special beds, failed in the application of their specific distinctions to the beds themselves, which they were meant to characterize. Lesquerox, standing on the mounds of waste at the mine's mouth, and studying his species with the rocks before him in which they lay, could not fail to discover the order of time in which the species made their first appearance, the groups

into which they formed themselves at each successive vegetation, and thus the surest answer to the constantly recurring question: Which bed is this? and this? It is very remarkable that of six or seven hundred species of coal plants already known, so large a number as four hundred are common to Europe and America. Among the lately described coal-plants of Hungary, Mr. Newberg recognizes two more of his undescribed species, which have been unknown in Western Europe but now appear upon its eastern limits. Of two hundred and twenty species Lesquerox found that one hundred were identical with those of Europe, and fifty more perhaps identical. The rest were closely related in their forms to European species; and the most common species there are commonest also here.

The fact of principal interest obtained by this preliminary study is, that the coal era opened apparently with large trees and closed with shrubs and ferns. At least, lepidodendra, calamites, and sigillaria crowd the lower sand-rocks with their vast trunks, the clays with their roots and the coal bed slates with their leaves,—while the upper slates and coals are full of fern bracts, and comparatively seldom show a tree. There are, however, reasons to believe this appearance a deception. If cataclysms swept the earlier conglomerates down from the shores into the sea, they would furnish to it multitudes of forest trees, which, when the times grew quieter, would remain upon the land unharmed to moulder where they fell. Or, if great river currents caused the sandstones, these would also be in the earlier unquiet times loaded with the plunder of the inland forest, but afterwards would embed nothing but a swamp vegetation of reeds and ferns. (Appendix VI.)

It is particularly desirable that private persons who collect what they call "handsome specimens," and transmit them to some distant public cabinet, would see the importance of labelling them not merely with their own names and that of the town in which they reside, but with a short and precise description of the exact place where the plant was found, on what stream, or hillside, how far above the nearest stream (not how far above *tide level;* that is worse than being righteous over much); but above all, if possible, and it is almost always possible, under or over what bed of coal, how many feet, and in what connection, whether in lime, in shales, or in sandstone; and if a small neat section of the hillside, with the place of its coals, and of the fossil when found, marked upon it, were added on the label, which might easily be done to the great advantage and improvement of the doer, its value would be perfect for the student, into whose hands it is sure at last to come for the accomplishment of its mission. Let such a system of collection be generally pursued, and our cabinets will soon cease to be curiosity shops and become workshops of science and factories of general truths in the highest and best sense those terms will bear.

21. **The Lower Coals** form in Western Pennsylvania a system by themselves, as has been said already. Clinging as it were to the face of the Conglomerate the lower system fared better than the upper one, and has been left to cover an immense area. In fact it forms by far the largest part—perhaps four-fifths of all the coal remaining on the surface. In Ohio (except near Wheeling), and in all the Western States it is the only coal, and may have been originally the only coal deposited. It may never have been covered by an upper system. It is possible that while the upper vegetation flourished at the east the western lands had become too high or dry to sustain in its vigor the coal vegetation; or had sunk too low beneath water level to receive anything but oceanic or lacustrine deposits of silt and lime.

22. Wherever the dip is gentle, this lower coal system prevails, the upper being swept away; but where the dip is steep, and in the middle of the narrow troughs, it receives the upper system on itself. It furnishes the beds of northern and western Pennsylvania as far south as the Conemaugh or Kiskiminetas, those of the Alleghany River and its branches, and of all the country northwestward of the Ohio. It occupies the west

and south of Virginia, and provides the coal of Kentucky and Tennessee. The cannel coal is perhaps exclusively of this system. Its beds were lettered, A, B, C, D, and E, in the diagram which the author prepared in 1840 for the fifth annual Report of the State Geologist, see secs. I., II., III., p. 81, and IV., V., VI., VII., p. 82. But such a nomenclature will not answer now. Numerous small coal seams, and even beds of slate neglected in this numbering, require notice and admission into a much larger and better discussed catalogue of beds. At that time, the local feeling of the party threw this lower system of five beds into pairs, A and B, with its underlying limestone; then C, D, and finally E, with its underlying limestone. A wider examination has not broken up this scheme of twin subdivision, rude as it is. The "Alleghany system," as the reports call it, shows the same, and on the Kenawha, in Western Virginia, the same is curiously well pronounced; see sec. XXXV., at Peytona. At that time also, a large bed in the upper part of the system was familiarly called the Elk Lick coal, from its locality near the romantic falls of that name in southern Somerset. This bed, which is the Upper Freeport bed of the Kiskiminetas and Alleghany Rivers, seems to be represented by the large upper coal of the Kenawha and Coal Rivers of Virginia (see XXXV.), and by the great bed at Karthause and Clearfield to the north (see VIII.). It marks the upper limit of the lower coal beds, and is covered at no great distance by the remarkable sandstone stratum hereafter to be described. This coal bed sometimes rivals the Pittsburg bed in size and purity of minerals, but wants its regularity. This is its fault, in common with all

Fig. 22.

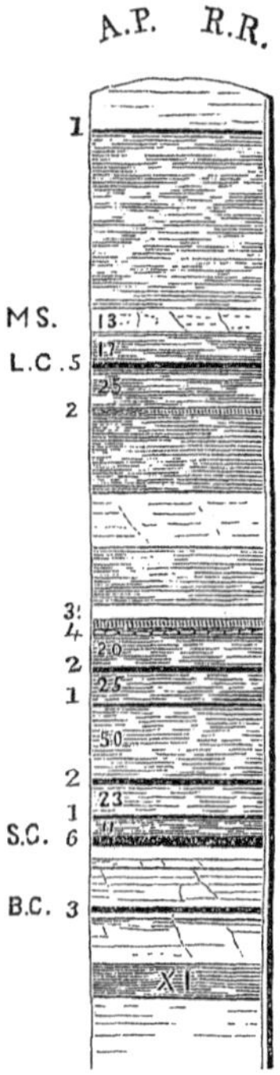

the beds of the lower system; they cannot hold their own for any great distance in any given direction. This is particularly true of the large bed B, which lies nearly upon the conglomerate (it has been called A in the 5th annual report), and seems coextensive with the coal field. At Towanda (T), (see Fig. 25, p. 120), on Broad Top (B, T), at Johnstown, on the Tennessee River (C), even at St. Louis (St. L), its sections can scarcely be told apart. Everywhere it is about fifty feet above the conglomerate; everywhere it has a small satellite some yards below it; everywhere it is itself a variable stratum, from five to twenty feet in thickness; a double bed, with an even roof and uneven floor, rising and falling stormily on a sea of fireclay, which sometimes has a depth of thirty feet.

23. A layer of cement stone, a fossiliferous limestone with the buhrstone iron ore upon it, and the beds of cannel coal, are the only other members of this system remaining to be mentioned with particularity; and these will be found included in the following abstract of the examinations made in Western Pennsylvania, chiefly by Mr. McKinney, and published in the fourth and fifth annual reports of the State Geologist. To say that it is a complete scheme of the lower series of coal measures, even upon its own ground, would be to ignore the enormous difficulties of a work covering so much wild and unbroken ground at that early day, when neither the waste of the forest, nor the manufacture of iron, nor the stimulus of speculation, had compelled farmers and land agents to prove the beds. To say that much has been done since to extend inquiry in a systematic way is also unfortunately impossible. Thousands of private drifts, it is true, have been begun, enough to expose every inch of the coal formation to an analytic scrutiny had the eye and hand been waiting to see and record. But these opportunities to science have been thrown

away in the absence of men whose official business it would have been to seize them. The drifts have each in turn fallen in, and future research must provide over again its ways and means of discovery. Nothing but a permanent bureau of geology, forming part of the organization of the State government, will remedy this evil; and nothing but the disgust with which the people of the State were taught at last to regard the survey under its old organization can explain the fact that year after year passes and no attempt is made to put the geological examination of the State on a true basis of cheap and perpetual action, not for the advantage of private persons, as such, but for the advantage of each and all the members of the commonwealth.

24. System of the Lower Coals.

	COAL (*Pittsburg*). 8 feet.	
	Limestone, k, Ore and shale	25 feet.
	Shale and sand	30 "
	Limestone, j, 4 ft.	
	Red shale	12 "
	Limestone, i, 4 ft.	
	Yellow shales	10 "
	Buff shales	18 "
BARREN MEASURES.	**Red** shales	4 "
	Limestone, h, 3 ft.	
	Shale and slaty sandstone	10 "
	Red marl	10 "
	Gray building stone	70 "
	Olive shales	100 "
	Limestone, g, 2 ft.	
	Coal, G, 1 ft.	
	Red and blue calc. marls	20 "
	Slaty sandstone	30 "
	Shale	thick
	Coal, F, 1 +	
MAHONING SANDSTONE, triple,		75 "

LOWER SERIES.	Shales		50 feet.
	Coal, E, (*Upper Freeport*)	6 +	
	Limestone, e,	8 ft.	
	Shales		50 "
	Coal, D, (*Lower Freeport*)	3 +	
	Sandstone with thin coal		70 "
	Shales with thin coal beds		100 "
	Coal, c, (Cannel).		
	Shales		25 "
	Buhrstone Ore,	1½ +	
	Limestone, c, (*encrinal*)	23 ft.	
	Shale		30 +
	Coal, B,	3½ ft.	
	Shales—(coal beds)		40 feet.
	Coal, A,	thin.	
CONGLOMERATE			15 "
Shale			50 "
CONGLOMERATE			30 "

---

25. *Description of the above System.*

Pittsburg coal.

Limestone, k, dark blue and black layers, 1—1½ feet thick, separated by shale, the main body of shale above and of limestone below; black layers ferruginous and bituminous. Shale full of iron ore along the base of Chestnut Ridge, in George and Union Townships, Fayette Co.

Ore, carbonate of iron, in 3 layers; first, 2 feet, second, 4 feet, third, 7 feet below the coal. At Brownsville and Connelsville a thin coal bed separates the two lower layers of ore, 25 ft.

Shale and Sand; gray slate, thin sands, soft shales; near Pittsburg a building stone; ripple marks, reeds; deposit universal, 30 "

Limestone, j, dark blue, weathering yellow ; hard, heavy, square, oblong debris along the Monongahela and Youghiogheny, 4 ft.

Red *and Yellow Shale,* 12 "

Limestone, i, thin, 2 "

Yellow Shale, purple below ; calc. nodules, 10 "

Buff Shales, 18 "

Red *and Blue Shales,* East Liberty and Pittsburg, 4 "

Limestone, h, lowest non-fossiliferous bed ; yellow and buff, hard, square, argillaceous, specked with crys. carb. lime, ferruginous, very heavy, 3 "

Shale Slaty Sandstone ; blue, yellowish green, stained, 10 "

Red *Marl,* blue, mottled, making a purple band along the hill-sides ; slightly calc. and fossil. It extends from the Alleghany Mountain in all directions northwestward and westward towards Ohio ; is sometimes 20 ft. thick, 10 "

Gray Building Stone, lower part solid, upper part slaty ; light yellowish gray ; brownish green ; splits into prisms ; sometimes micaceous ; sometimes weathering badly and crumbling ; finely false-bedded ; at the bottom many pebbles of blue or drab fine clay full of specks of mica, from 1 to 15 inches diameter, round and water-worn [?]. Lower part shows traces of coal bedding, bituminous slates, interlaced carbonized plants. Where the rock has much iron, its white quartz grains crumble out ; small univalve shells in the green shales ; sometimes the whole is a mass of loose sand, with coarse nodules of carb. of iron. The stratum forms cliffs up the Monongahela, until it sinks below water level, and spreads up over the Northern country, 70 "

Olive and Buff Slates and Shales ; nodules of limestone ; slightly micaceous ; sandy ; fissile ; occasional thin layers of heavy sand-rock ; fine impressions of plants in the lower dark slates ; exposed throughout Alleghany County and in the countries around. [It is in fact universal], 100 "

Limestone, g, Upper Fossiliferous, dark gray, black, hard, fossils numerous but not various, productus, leptœna, terebratula, Encrinus (sometimes half an inch in diameter), orthoceras (abundant, from ½ to 2 inches), ammonites. Stratum wide spread, 2 "

Coal, O, 6 to 18 inches, good, brilliant, hard, 1 "

Red *and Blue Calcareous Marl,* just above the Ohio at Pittsburg ; mottled, soft, calcareous, no regular cleavage, homogeneous, friable, weathering easily; ferruginous limestone nodules, sometimes in whole beds; a few small fossils; exposed up the Alleghany to Sharpsburg, and on Thompson's Run, [20?] ft.

Slaty Sandstone; light gray, thin, variable, quarried, forms the bed of Pine, Deer, Sandy, Plum, and other Runs up to Tarentum; at Freeport and Kittanning on top of the highest hills, 30 "

Shale, yellowish and brownish; thickness considerable.

Coal, F, 1 foot, 15 miles above Pittsburg, Kittanning, Alleghany furnace, 2 ft.; near Butler, &c., 1 "

[**MAHONING SANDSTONE**]; triple bed, two layers of sand, each 35 feet, and one of shale, 25 feet; sandstone very coarse, compact, heavy, sometimes 4 feet of round water-worn white quartz **conglomerate** below the upper layer, containing nodules of iron ore; sometimes below the conglomerate 6 inches of **coal;** such is its character as it ascends the hills slowly from Tarentum to Freeport, and forms an east and west belt of high hills through Buffalo T., Armstrong co. Near Kittanning its triple character is lost. It nearly reaches the Red Bank on the highest knobs. In Butler it covers the ground between Buffalo Creek and Slippery Rock, through Centre, Butler, Middlesex, Muddy Creek, Slippery Rock, Parker, and Donegal townships, past Porterville, and appears on the western side of Mercer County, 75 "

[This, however, is a meagre account of this important stratum which reinaugurates the Conglomerates, and should be called our **Fifth Great Sand-rock.** It would, in fact, be as correct a basis for the upper system of coals as the Conglomerate is for the lower, and then the Barren measures would become an unnecessary term. It is as widely spread as we have opportunity to examine it over the Palæozoic ground, although on account of its high position it has suffered immensely more by denudation than its congener below. Its triple structure is its most curious feature. This, so far from being lost in going only from Freeport to Kittanning, is a characteristic as far to the eastward as the Juniata River. The description above given of it at Freeport will answer for it almost without change on Broad Top in Bedford County, where the conglomerated white quartz pebbles are again seen at the base of its upper stratum, and a **coal bed** (thickened of course eastward to 5 feet) lies between its parts. It has there received the sobriquet of the "Top Rock." Forty miles northwestward on the head waters of Clearfield Creek it cumbers the meadows with cyclopean blocks as large as cabins; and near Latrobe its outcrop lies in a solid plate twenty feet thick, upon the gentle slope of a hill, down which vast quadrangular masses have been slid by the denuding wave from a few inches to a hundred yards, like ships upon their ways. In the South it is as persistent as in the North, and may be recognized in cliffs upon the Kenawha and Big Sandy as well as upon the Susquehanna. The author offers therefore, with some confidence, the identification of this rock with the remarkable conglomerate above the twelfth coal bed in the Shamokin Basin and its representatives in the other anthracite basins.

Shale; brown and blackish, laminated, friable, from 2 to 20 ft. at Freeport; but in Butler 50 ft.

Coal, E, [*Upper Freeport*], rich, compact, inflammable, cokes well, sometimes pyritous, on the whole of equal quality with the Pittsburg bed [makes 4.8 cubic feet of gas to the lb.] emerges on the Alleghany 3 miles below Tarentum, 6 ft. thick; at Freeport 144 ft. above water, 4 ft. thick; at Kittanning 370 ft. above the water; on Mahoning and Red Bank and Sugar Creek in the highest summits of the upland. It is everywhere known and wrought back from the river westward through the central parts of Butler (4 ft. thick), and crosses the Beaver River. Eastward it rises and falls over the anticlinals which cross the Kiskiminetas [and makes its appearance in the Ligonier Valley and in Somerset County as already described]. Westward it sinks beneath the Ohio above Wheeling, and reappears upon the waters of Western Virginia as the large top coal.

Limestone, e, separated from the coal by fireclay or blackslate; sometimes a bed of nodules in shale; yields ash colored lime; sometimes very hard, dark blue and pure; fossils few, and chiefly in the shale above—thickness, 6 to 8 "

Slaty Sand and Shale. 50 "

Coal, D [*Lower Freeport*], 2 to 4 feet thick, 57 ft. above water at Freeport; on Beaver River it is a bed of coal of tolerable purity and great regularity, about 3 feet thick, on which the coal trade of the River has hitherto chiefly depended. [It is a bed coextensive apparently with the Coal Field, appearing at Johnstown, at the Portage as probably the Lemon's Coal, and on Broad Top as the Cook Vein, the finest bed upon the mountain, 8 feet thick, with a thin undermining of slate which occasionally turns to rock, and sometimes disappears.]

Sandstone; upper part sometimes shale, coarse compact gray, and brown layers, contorted and full of sigillaria, lepidodendra, calamites, &c., the first in great abundance [see opposite N. Brighton]; many thin bands of shale and irregular lenticular deposits of **coal** from 1 to 2 feet thick, thinning out in a few yards, 70 ft.

Shale, soft, carbonaceous, embracing occasional layers of sandstone and fireclay, and **two beds of coal** from 1 to 1½ feet thick. [Near Mercer one of these beds is 5 feet thick]. 75 to 100 "

Coal, C, [*Kittanning*], **Cannel**, emerges from Alleghany 6 miles above the Kiskiminetas; at Kittanning is 6 feet above high water and 3½ feet thick; contains pyrites; is easily traced to the mouth of Clarion. On the Cowanshannock, Crooked and Red Bank Creeks it nearly reaches the Indiana and Jefferson county lines; it is worked at Horner's Mills on Buffalo, and crosses Conequenessing and Beaver [280 feet above the seventh level; if this be not one of the beds of coal mentioned in the shale above. It is there a peculiarly dry light coal analyzing 68 carbon, 35 gas, 2 ash, and .16 sulphur]. The Shales beneath this coal sometimes disappear and let it down upon the limestone, shutting out the iron ore next to be described.

[This bed is destined to be celebrated in the future coal trade of this country as the great depositary of Cannel]. On the south side of Red Bank Creek in Red Bank T., Armstrong Co., Mr. Alex. Cathcart opened a bed of bituminous coal 2 feet thick, supporting 8½ feet of light, compact, lustreless, conchoidal slaty cannel coal, igniting freely, burning brightly and leaving a moderate amount of ash. The cannel coal is not traceable beyond this neighborhood, being merely a local modification of the deposit. Three miles from Greensburg, Beaver County, it occurs again under the same conditions, eleven feet thick. [The author visited this mine during the survey, and traced this bed in its bituminous and slaty forms over all the northern part of Pennsylvania. Since then he has seen it show the tendency to take the Cannel form in several places, and can prove its identity with the Cambria Co., the Darlington and the Peytona Cannels. What seemed at first a local affection proper perhaps to several or all the coal beds of the west, must now pretty certainly be regarded as a characteristic of this Kittanning Coal Bed either alone, or in common with the bed next over it; for on Coal River in Virginia the next above it yields some cannel also. Curious specimens may sometimes be obtained which refute also the opinion that the bituminous and cannel parts are not at all related, or related in unalterable super and sub-position, by showing their *alternations* like those of slate and sand, or slate and coal. There can be little doubt that Cannel coal is produced by an infusion of resin in bituminous shale, and this explains on botanical principles the fact of its occurring only or principally at this stage of the Coal Formations, at which time only we may thereby conjecture those plants grew which yielded this unknown resin.

From these localities—6 M. E. of Franklin, Venango Co., from Conequenessing, and from Darlington (Greersburg), three analyses gave the following results:—

Volatile matter 43.	Carbon 37.	Ash 20.
" " 33.	" 51.	" 16.
" " 24.	" 42.	" 34.

Buhrstone Ore Shale; brown and black with nodules of carb. of iron, and layers of sandstone, sometimes as a solid stratum 10 feet thick in the middle of the shales; thickness 25 f

Buhrstone and Ore; a bed of hard, gray, yellowish chert or flint stone, cellular and worm-eaten from the weathering out of iron and lime. The ore lies on the buhrstone and under the shales; is a brown peroxide at the outcrop, and protocarbonate under cover. Where the chert abounds, the ore is lean; when the shales above are free from sandstone, the ore is thick and good. [Both chert and ore are deposits subsequent to the deposition of the shale which contained them, but previous to the denudatian of the country; for the ore occurs sometimes as thick at the outcrop as where it has the full deposit of shale above it. The leaching process which carried off the iron and silex disseminated through the shales down upon the face of the calcareous mud must have found the latter an unbroken impervious plane, and not, as now, rent in fissures through which the waters find their way with such ease that gangways in the ore are always dry. This is another datum we have for calculating the date of the Denudation. It is this leaching process which has converted the carbonate into the oxide and thus enhanced the theoretical percentage of iron in the ore from 46 p. c. to 56 p. c. An analysis of a specimen from Armstrong Co. is given in the 4th annual report of Mr. Rogers, to this effect:—

Carbonate of iron,	68.32.	
Carbonate of lime,	15.54.	
Insoluble (mud and sand),	10.58.	
Iron.		32.95.

Water, 4; Carb. mang., 1.35; Traces of magn, and mang.

Another specimen was to this effect:—

Carbonate of iron,	54.33.	
Insoluble,	40.90.	
Iron.		25.34.

Water, 4; Lime, magn., alum, traces.

Another from Warne's, on Bennett's Branch, Clearfield Co.

Carbonate of iron,	55.10.	
Peroxide of iron,	9.50.	
Carbonate lime,	5.80.	
Carbonate magnesia,	5.40.	
Insoluble; sand and mud,	21.00.	
Iron.		34.72.

Water, 3.

There remain in the shales above the ore plate usually a vast number of nodules of carbonate of iron, more or less mixed with silex and lime, and these are often in sufficient proximity to admit of mining with the ore. In places, it is beautifully variegated with the disks of encrinites crystallized white upon a blue and purple ground, and sometimes the iron itself is crystalloid. Its surfaces are frequently mamillary.

The ore bed proper varies from an inch to 5 feet, rapidly changing its form, composition and thickness at every step. It yields best wherever there is a bowl in the limestone. In the openings two miles south of Shippensville in Clarion Co., it has reached 9 feet, throughout which the silex is disseminated.

[Its best average may be stated at less than 18 inches, and a common average over large areas at 10 inches.]

Fossiliferous Limestone, c, compact, light, blue, disposed to weather into thin layers, but very hard, sometimes slaty, always interrupted vertically by siliceous beds, burns gray, abounds in *encrinal disks* of crystallized carbonate of lime, innumerable small shells (*terebratula*), with occasional sharks' teeth. It varies from 10 to 20 feet, and has a very wide known range.

It rises from beneath the Alleghany River about 4 miles above the mouth of the Kiskiminetas (where it is 100 feet under water), and slowly ascends the river and the hillsides together, outcropping on all the tributary valleys and finally throwing its waving outcrop across the highland in ranges of low summits covered with chestnut and oak, and distinguishable from a great distance in a landscape filled with hemlocks and pines. These knobs cross through Pinegrove, Elk Creek, and Irvine townships, in Venango. At Kittanning it is 100 feet above the river; and several hundred at the mouth of Bear Creek; it occurs on Red Bank east of N. Bethlehem, and ascends both sides of Clarion [and has been traced through Elk Co. into McKean, along the centre of the 4th basin. It is recognizable as far east at least as Karthaus and Clearfield; and southeast throughout Indiana, Westmoreland, Cambria and Somerset into Virginia. It is not everywhere accompanied by the buhrstone, but almost always has over it traces of the ore more or less remarkable. Throughout the south and east it scarcely can be said to exceed 4 feet, and seldom shows any peroxide except at the crop.] West of Kittanning, on Buffalo Creek it is 15 feet thick; it comes within 10 miles of Franklin, and shows itself on Bear and Sugar Creeks; emerging on Slippery Rock and Beaver River, it is 20 feet thick at the mouth of the Conequenessing, 150 feet above the water; is wrought near Centreville and Harrisville; is high above the Neshannock at Newcastle, a compact blue rock ten feet thick; shows a singular shrivelled face of equal thickness nine miles higher up; dwindles to a dark bituminous fossiliferous slate 15 inches thick at Sandy Lake, and ranges its outcrop thence past Mercer, where it is 2 or 3 feet thick, into Ohio. It descends the Beaver River Hills slowly toward water level past Brighton, and goes under the Ohio, 20 feet thick, some miles below Beaver. It cannot be recognized on the West Virginia waters. Average thickness, 15 ft.

Shale, sometimes thickened with beds of sandstone; contains much nodular **iron ore**; at Alleghany Furnace 2 miles above Kittanning nine layers in 12 feet; has yielded well at Scrubgrass and Rockland furnaces; [occurs at Johnstown and elsewhere.] 20 to 40 "

Coal, B, good; 3½ feet at Leonards above Kittanning, over 3 feet fireclay; 233 feet above the Alleghany at Robinson's salt works, nine miles below; opened frequently southeast of the Clarion; 5 feet thick west of the Alleghany, over which are three thin coals; ranges through Irwin and Sandy (Venango) 12 feet below the limestone, 3½ feet thick with a rider coal of 20 inches in the shale above; 30 feet below limestone at Lucinda Furnace, Pinegrove (Clarion); at Mercer lies just under the limestone 3 feet thick; also at the mouth of Beaver River, 4 feet thick.

[I have little doubt that this is the great bed (A) near the bottom of the coal measures, above Johnstown just under water level; Shoenberger's coal at the foot of Plane No. 6, Portage R. R., the Tunnel Bed and Bell's Vein (Alleghany mt.); the Barnet, Crawford, Riddlesburg, Hopewell, Loy and Patterson Bed of Broad Top; the large Tangastootac, Queen's Run, and Heron's coal, the Ralston, Blossburg, and Towanda bed (B) of the northeast. Everywhere it has one or two small riders, and rests upon an irregular mass of fireclay as before described, p. 93.

Shale and mud rock; upper part containing nodular carbonate, and sometimes much sulphuret of iron; lower part *contains thin coals* and bituminous slates, three to ten inches thick, 40 "

Coal, A, lowest workable bed on Alleghany River; yields 20 inches semi-cannel, light, slaty, peculiar lustre at Brandon's, 6 m. E. of Franklin; at How & Bratton's (Rockland T.), yields coal 4, 3 and 18 inches, separated by 12, and 36 inches of blackslate, making two feet of coal in six of gangway; it is traceable in Irvine (Venango).

XII. **White Sandstone, No. XII.**, base of the coal measures, coarse, massive, in two solid beds, upper, 15, lower, 30 feet thick, with shaly sands between, with nodular iron ore; consists of minute grains of quartz so loosely held together as to fall into pure white, dry sand, supporting a stunted vegetation over extensive glade-lands on the north borders of the coal basins towards Lake Erie and the head-waters of the Sinnemahoning. [The characteristic marking of the surface is a circular blotch of brown or blackish color, attacked by the weather, and resulting in a small pit; the whole rock cliff is sometimes pitted, and at other times roughened by a network of ridges from one to six inches high, as described on page 75]. Furnace wall stones are obtained from the soft light gray variety of this formation, which is easily dressed, and resists the fire well. No. XII. caps the hill country on both sides of the Alleghany River near Franklin, and forms the general upland of Northern Venango. As it rises, the whole country rises to an elevation a thousand feet above Lake Erie, and its last escarpment breaks off within the New York state line, when its whole topographical aspect is peculiarly characteristic. On Teonista Creek one of its members, 8 feet thick, lies about 50 feet above the water, and is a true conglomerate, with rounded pebbles of quartz. Two miles south of Tytiute, it caps the hills with a coarse conglomerate, traceable to Warren, up French Cr. to Meadville, and down the Shenango to Sharon and Newcastle [disappearing below the Beaver River dam, four miles above New Brighton; the large coal above it is in the lock chamber].

In S. Ohio, as for instance, along the Salt Creek branch of Hocking, the conglomerate proper is supported by two hundred feet of fine grained sandstone in cliffs facing the deep and narrow ravines filled with evergreen forest, and offering unexcelled beauties to the landscape painter (Briggs' Report, Ohio, 1838). Mr. Foster justly grows eloquent about the overhanging cliffs of the Licking River, with cool and spacious grottos underneath, and waterfalls three or four score feet in plunge. "Nothing can be more grateful than one of these retreats during the sultriness of summer." The Indians felt the charming influences of scenery like this, and carved on the isolated masses of the rock their rude symbols of history or religion. One of these has been destroyed in making the Ohio Canal; it bore an outspread hand, and the place is still called "the rock of the black hand"—(p. 99.) The rock is here an agglutination of coarse white sand, with quartzose pebbles, and Mr. Foster supposes it a deposit of the estuary of a great river, with ripple marks and indistinct fucoides.

26. It would extend this book beyond the author's present purpose, to give similar abstracts of reports of geologists in other parts of the United States, a work, by the way, much wanted, from some judicious hand; but the following, from the Ohio reports of 1838, will serve to correct and enlarge the above, as well as to give some promise of what may be expected from a careful survey of that State, if one should ever be ordered and intrusted to competent men, upon a scale befitting its wealth and relative importance in the Union.

The **Ohio Coal Series,** *west of the Muskingum*, were described by Dr. S. P. Hildreth in the Report of 1838, but no lower than the buhrstone, thus:—

34. *Slaty Sand-rock*, argillaceous; "Barris's Grotto." 80 ft.
33. **Red** *Shale;* universal; kidney ore, semihæmatite, sp. gr. 416. Found from Shade River to Marietta; in Olive and Lebanon (Meigs), several feet thick.
32. *Mica Sand-rock*, slaty; crumbling; hilltops, 40 "
31. *Ochrey Shale;* nodules of ore; 4 "
30. *Sandstone*, fine, light blue-gray; divided by occasional micaceous layers; makes good grindstones, (Warren T.) 25 "
29. *Arg. Sandstone;* fine, hard, upper part slaty; quarried frequently; more valuable because containing no fossil plants, which are found in all the lower sands of the series. 25 "
28. **CAVERNOUS SAND-ROCK** [**Muskingum Conglomerate**]; 60 "

Conglomerate, containing fragments of coal, carbonized wood and madrepores, and rounded lumps of arg. sand and slate, at Barris's Grotto, below the mouth of Hockhocking, a loose puddingstone, with very small pebbles and coarse gravel, cemented by tufaceous lime, which falls out in thin layers under the weather, 18 feet; above, compact, coarse sandstone, massively bedded and cleft, throwing down vast cubical fragments to the base of the cliffs; an excellent and durable building stone; above, sometimes a layer of coarse, sharp sandstone; quarried at Marietta; cliffs 60 feet high between the Hockhocking and Muskingum and up the Muskingum; high hills, mouth of Shade; covers Washington, part of Morgan, Athens, Meigs.

27. **Coal,** 1½ ft.
poor, slaty, sometimes a few inches thick; traced from South Wolf Creek to Duck Creek.

26. *Sandstone*, micaceous, compact below, slaty above. (Roxbury T.) 20 "

25. *Marls;* gray, **red**, green,
Sandstone, schistose mic. (a few feet), } 20 "
Upper gray marl has the most lime; the middle brown marl decomposes into loam; the lower green into a potter's earth; mostly clay marls; sometimes a limestone; ash marls contain shells; fossils, plant stems and leaves, mostly ferns and calamites. (Grotto of plants, two miles below Marietta; Barris's Grotto below Big Hockhocking.) Best developed at Fairchild's Mill, Decatur (Washington); extends across Barlow, Wesley, Waterton, Waterford, into Athens; and south to Gallia; also in Wood, Va., and both sides of the Ohio from the Guyandotte to the Fishing Creek. The *Yellow Pine*, once abundant over this region, is now restricted to these chocolate colored shales, which produces also the best wheat, corn, and grass soil. When the upper sandstones come in the yellow oak, chestnut and poplar take the field (head waters of Moxahala, Sunday, Federal, Wolf Creeks).

24. **Limestone** (non-fossiliferous), 40 "
Stratified, 1–4 feet; upper and middle portions buff, gray, blue; parted by marls 3–4 feet thick, pale blue or dark-brown; buff, prismatic, weathering; blue, compact, conchoidal. Lower part dark; sulph. iron in cubes. Best developed above McConnelstown, 50 feet; well seen in main

Wolf Creek; creek bed for nine miles above its mouth. One mile east of McC. 250 feet above Muskingum; at Coal Run mouth it is just above water level; goes under at Doval's ripple, 5 miles up the Muskingum; traced eastward on Meigs, Olive, Greer, Duck Creek, and Little Muskingum (Morgan and Monroe); westward nearly to Sunday Creek; southward down the Hockhocking nearly to its mouth.

23. *Sandstone*, argil; slaty, somewhat mic., 12 ft. }
Sandstone, hard blue, 1 ft. } 18 ft.
Blackslate, changing to ash, 8 ft. }

22. **Coal**, 4 "

"Limestone Coal;" slaty and thin on the Ohio; good in most places. McConnelsville, 250 feet above river; at mouth of Wolf Creek (Washington) at water level, 12 miles east of McC. (dip 20 ft. to mile); Meigs, Olive, Green, and Duck Creeks; Coal Run (S. W. corner of Roxbury T.) 5 feet thick (1 ft. coal, 1 ft. slate, 4 ft. coal); west branches of Little Hocking (Decatur T.), 3–4 ft. thick; runs out on the hills of Federal Creek; on the Ohio, thin base of cliffs.

21. *Fireclay*, 3 "

20. **Limestone**, 6 "

Hard, tough, argillaceous, dirty gray; no fossils.

19. *Blue Clay; no mica; no sand*, }
Coal, one inch, } 8 "

18. **Red** *Shale*, calcareous; layers of loose yellow limestone, 50 "

17. **MAHONING SAND-ROCK**, coarse, friable, massive; cliffs, 52 "

Heads of Shade River, 80–100 feet thick; between Lodi and the Ohio, 40–50 feet; at Pomeroy, its upper part contains a layer of good non-micaceous drop stone; at Dr. Martin's, on Muskingum, same, covering 2 ft. of **Coal**.

16. *Shales; iron ore;* cemented ferrug. fragments; best seen in Cheshire T. (Gallia), Secs. 19, 20, 25, 2 feet thick, cubical yellow oxide; overlaid with coarse sandstone.

15. **Pomeroy Coal**, 5 "

14. *Fireclay*, 3 "

13. **Limestone**, 2 "

Its west outcrop ranges from South Morgan across Athens and Meigs to the Ohio River; well developed on Federal Cr. West Shade River, Leading Cr., Ohio River. Dips E. S. E. except just south of Athens, where is a fault or axis. Below Kuyger Cr., Addison T. (Gallia), it is 1½ feet thick roofed with bit. shale 3 ft. 150 above the Ohio; 5 miles up the river it is 4 ft., 70 ft. above water. Shows the limestone below; at Leading Cr. mouth, 5 ft., 40 ft. above water; near Pomeroy, 14 miles above Kuyger Cr. (7 miles due E.), it goes under water. Up Leading Cr., 7 miles, 6 ft. thick, 150 ft. above creek level; dip east; *sand-rock directly on coal.* Runs out in Gallia and S. W. Athens, growing thin. Going N. from Pomeroy, it occurs in Car's Run bed, at one mile. Reappears on Thomas's Fork of Leading, 5 ft.; shale roof. Reappears on W. Br. of Shade, 4 ft. thick; clay floor and roof; clay with nod. limest. and sulph. iron; *on roof clay*, 6 *feet of bituminous shale burning freely:* over all 50 ft. of sandstone. Near north line of Meigs, has 10 feet of shale above it with nod. lime, *over which is a thin coal*, and then the sandstone. In Adams, it reappears on Pratt's Fork, and rises towards Athens, where the sand-rock over it becomes very great and makes a hilly country. On Federal Creek, a little above the mouth, 5 ft. thick, goes under water. Can be traced up the creek to Marion Township, where the Upper Foss. Lime. lies 80–100 below it; here (30 miles N. of Pomeroy), coal 4 ft., clay floor; roof ash shale 1 ft., rich bitum. shale 1½, thin coal 1½, massive coarse-grained sand-rock.

In roof shale 20 species of plants, equisitaceæ, filices numerous; lycopodiaceæ rare. In coarse sand-rock, multitudes of tree stems 4–10 inch. thick, 2–3 ft. diameter, compressed, one side concave, other convex, siliceous casts full of worm holes (extreme N. branch of Shade River, Lodi T., Athens Co., and generally the runs south of Athens).

The limestone is best developed 4 ft. thick, sec. 24, Town. 7, Range 12, Marion T., Athens Co., head-waters of Federal, but always attend the coal; sometimes nodular, sparse, or thin, no fossils.

12. **Shales,** *Clays, Slaty Sand-rock,* 80 ft.

11. **Limestone.** *Upper Fossiliferous,* 8 feet.

In regular layers 6–12 in. sometimes separated by calc. shales, increasing the whole to 12–15 ft. Sometimes the bottom layer is a conglomerate, of fragments of limestone small as a pea, rounded and polished. (3 miles S. of Athens; Morgan T. Gallia; Will's Creek S. W. corner of Guernsey) Everywhere full of encrini (joints) and very small terebratulæ, producti, gryphea, and few spiriferi; many equivalve bivalves and turbinated univalves not known below. At Sharp's Fork of Federal, 15 ft. thick; crosses West Meigs into Gallia; head of Campaign Creek, lower layer black, on coal; traced down Chickamoga Cr. and sinks near Gallipolis below creek water; base of hills at Limestone Run, 8 m. north; sinks on Leading Creek at Upper Salt Well; underlies the Ohio River for many miles.

10. *Shales*, slaty, capped sometimes with thin coal, 30 ft.

9. *Sand-rock*, compact, coarse, siliceous; calamites, 25 "

8. *Shale*, clay slate, 10 "

7. **Coal**, good 2 feet.

6. *Shale*, bituminous; nod. ore; 15 "

Upper part slaty, smooth sandstone.

5. **Coal**, slaty, 2 feet.

Poor, light, cakes readily: near Zanesville 3 ft. 8 miles above McConnelsville, 20 inches. ☞ This is the locality of all the measurements from the 4th fossil. lime. to the calc. silic. rock.

4. *Shale*, slaty clay, light ash color; *fireclay* above. 15 "

3. *Sandstone*, coarse, loose, brown, crumbling, giving 20 " the sand beds to the Raccoon waters; seen well over the Buhrstone at Wild Cat's Den, sec. 26, Elk T.; more slaty on Muskingum; siliceous on Hocking.

2. **Iron Ore**, ½ foot.

Thin, brown, oxide, porous, mammillary cavities.

1. **Buhrstone** 9 feet.

"Calcareo-siliceous deposit;" of uniform texture; pure quartz; free from lime and oxide of iron; light gray; cellular; sound metallic; intensely hard; stratified and regularly cleft; the bed plane most cellular and used as the face of the millstone; more or less marine shells; interpolated layers of limestone 2 + feet thick above and below the buhr, which in such cases is but 2–4 ft. thick; greatest thickness in any one bed, 9 feet. The best Paris blue buhrstone, six and a half feet in diameter, sold in Cuvier's day for $240. The Ohio buhr was first wrought by Abram Neisby in 1807, and during the embargo superseded the French. From 1814 to 1820, 4½ feet stones sold for $370 a pair, 7 feet for $500; in 1834 4 ft. stones for $150. The rock is a mine of wealth to Richland, Elk, and Clinton Ts. and Hopewell Ts. Muskingum. It ranges from the Ohio River to Stark Co. and on northeastward, in a band from 12 to 20 miles wide with an easterly dip. At Toppin's Mill, Margaret Co. 6 ft. exposed, layers of 6 inches, gray, calcareous, splits smooth into window sills, &c.—Beds the streams in Lee T. roof loose sandstone, floor, dark shale. Further west a bed of coal occurs a few feet under it; still further west *it rests on coal.*—At Judge Warner's, Lee T. in road, 8 ft. thick, strata 8–10 in., upper layers calc. lower pure quartz and hornstone; black, green, blue, horn color; frac. conchoidal; lowest layer nearly black; top layers cellular; *this feature seems confined to its western outcrop.*—Seldom seen large until Elk T. 24 m. S. W. of Athens, where are many quarries.—Crops out on the highest hills of Richland T. (Jackson) 8 miles W. of McArthurstown. At Redfearn's E. br. mid fork of Salt, a conglomerate of water worn quartz pebbles cemented with sand and iron; covers the hills to the base with its fragments;—traced N. to Raccoon Cr. heads and Honey fork of Queer (Hocking);—southward rich buhr quarries in a belt 12–14 miles wide, 6–8 broad.—Middle of Wilkes T. (Gallia) crops under the bridge, 4 m. W. of Wilkesville; appears no more to the eastward; cut in salt wells on Leading and Chickamoga Creeks. S. W. it crosses west end of Gallia Co. head of Symmes' Cr. to Ohio River.—N. E. from Jackson Co. crops out E. side of Hocking Co. and York T. Athens Co. 8–9 ft. thick, roof coarse sandstone. E. of Hockhocking River, few fragments of it have survived weathering; N. E. corner of Green T. and in Monday Cr. T. seen in place; abundant in Perry Co.; continuous on Rush Cr. Pike T. Lexington, a pure quartz, used by the Indians for arrow heads; bears N. E. into the corners of Licking and Muskingum Counties, forming the "Flint Ridge" summits; in both Hopewell Townships it covers the surface with fragments; used by Indians for knives and spears; innumerable pits from Jackson to Muskingum.—In Muskingum color is lighter; no open fissures, but full of tortuous vermiform passages $\frac{1}{16}$ in. diam., the matrices of a fusiform univalve, and encrinal joints, with some terebratulæ, spiriferi, producti, &c. 8–9 ft. thick.—From Hopewell to mouth of Licking, dips 10 ft. to the mile.—S. along the Moxahala Creek lies high on the hills, yellow, soft full of terebratula; in York T. (Morgan) 8–9 ft. thick; traced down Island and Oil Runs to Muskingum River, and goes under 2 miles above McConnelsville.—At McC. bored through 110 ft. below water, *the lower or main salt rock lying* 650 *feet below it, with little variation, for* 10 *to* 12 *miles below.*—At Campbell's Mills 10 ft., floor bit. shale; some of it pure conchoidal limestone.

27. The resemblance, in fact the identity by this description and the one given before of the same formations is perfectly evident. To. Dr. Hildreth belongs the honor of having been the first to express the true spirit of the Ohio geology. Mr. Briggs took up the rocks thus characterized and reported their extent through Athens and Hocking Counties, in the Second Annual Report of the State Geologist of Ohio for the following year. He describes the Mahoning sand-rock as a true conglomerate forming mural cliffs between Athens and the Ohio;—the red shales of the barren measures as visible along the slopes of Federal Creek; the upper non-fossiliferous limestone as 30–40 feet thick on the hills below its mouth; the next 15–20 feet thick at water level just above its mouth; the Federal Creek or Pomeroy coal as from 4 to 9 feet thick; and thus completes the series *downwards* from the buhrstone:—1. The Dover coal 3–4 ft.; 2. The Nelsonville coal 5–9 feet, valuable, finely laminated, impressions between the laminæ; contains lenses of sulp. iron; roof sandstone; sinks beneath the Hocking 5 miles below Nelsonville. A few feet under it on Raccoon Creek is a heavy compact, bluish iron ore, the most valuable in that region, impressed with ferns, resting in a solid plate from 6 to 10 inches thick on shale, and covered with shales full of ball ore. 3. The lowest coal 80–90 feet below the last, 2–4 feet thick, laminated.

28. In Muskingum and Licking Counties Mr. Foster describes the same system of rock in similar terms.

Sandstone, mica: 40; dark shale 2; sand fossil 20,	62 ft.
Limestone, marly, 8; tough, 2; sand: thin, 6; lime: chalky, 1	17 "
Sandstone,	40 "
Coal 1; shale 1; **Coal** 3.6-clay 10,	15 "
Limestone,	4 "

Coal, Cannel, 2 ft. Grummin's Tavern, described in Sill. *Journal,* vol. xviii. 29, as the heaviest coal then known (1.6) conchoidal, resinous, brilliant; elsewhere poor and penetrated by Calc. Spar.

Coal; 2½;

Coal, 5–7 feet; the great bed of the country free-burning and sulphury.

Then follows the Zanesville section thus :—

Sandstone 40; blue slate 5:		45 ft.
Coal, 2–8, larger elsewhere. Bitum. shale 2	5 ft.	
Sandstone 40; shale 8		48 "
Coal, larger elsewhere,	2½ ft.	
Shale and Sandstone		25 "
Limestone, blue	4 ft.	
Blackslate with laminæ of **coal**	6 ft.	
Sandstone, fossil,		5 "
Blackslate with coal 3; **coal** 2-2, poor	5 ft.	
Shale blue, 6; sand: slaty 12; shale 7; sand: compact 6: shale with nod. iron ore 3; sand: gray, micaceous, 2; shale with nod. iron ore, 2½,		38 "
Blackslate,	3 ft.	
Sandstone.		6 "
"**Cannel Coal**" slaty, resinous, much ash	1-4 ft.	
Shale		5 "
Limestone, blue.	8 ft.	
Coal,—bed of the Muskingum River.	6 ft.	

In Hopewell Town these last rocks read thus :—

Limestone, fossil, cherty, 10—14 feet.

Cannel Coal 1; shale calcareous 5; cannel coal 3-5.

This is the only *workable* coal bed between the Buhrstone and the Conglomerate.

Mr. Foster describes the Buhrstone thus :—

The Buhrstone passes into hornstone, but is usually a grayish white siliceous chemical precipitate, cellular like amygdaloidal trap, full of infusoria moulds, drusy, with quartz in the oblique fractures, sometimes in six-sided pyramids, sometimes smoky, cavities sometimes filled with chalcedony; calcspar and heavy spar or sulphate of barytes; full of terebratulas, encrini, authophylla, spirifers, producti.

In the following section we see the Buhrstone deposit *repeated* locally in the series; which was to be expected, since its origin is no doubt to be referred to submarine siliceous hot springs.

Buhrstone 4; shale 10; Hornstone 1 to 4;		20 ft.
Limestone gray and cherty	5 ft.	
Shale, dark 30; light blue 10		40 "
Coal,	8 ft.	
Shale, light blue, 10; slaty sand, 8; shale yellow, 15,		33 "
Iron ore, 8 feet; dark shale, 10; iron ore, 1—4,		20 "
Limestone, brown, 5; light blue, 6,	11 ft.	
Sandstone, compact,		40 "

At least nine limestones are enumerated in the series in this part of Ohio; the highest is non-fossiliferous 3 feet; a hundred feet below it, is a second full of shells, 2 feet; below it, is a third non-fossiliferous, 6 feet; a fourth impure bed beneath the Jackson coal; a fifth on the top of Putnam hill, thin, buff, and fossiliferous; sixty feet lower is a sixth, with *encrini two feet long*, by which it may be recognized over all the country, 4 feet; a seventh, dark blue, and of a quadrangular fracture of extreme regularity, lies in the Muskingum river bed, and is known by its large *uniones*, beautifully preserved; a light blue, slightly fossiliferous limestone crosses the national road near Kent's run, and is 14 feet thick; the ninth, near Brownsville, is fissile and cherty, crumbles, and shows marine shells.

29. From Mr. Whittlesey's Report on Trumbull and Portage County, I take the following account of the coals on their extreme northern outcrop, chiefly for the purpose of showing how irregular the lower beds are, and how difficult it is to form any scheme of them which should answer at a distance. "The

strata are subject to continual distortions, forming basin-shaped cavities that sometimes sink 20 feet in a less number of rods." At Brookfield, where there is an area of 500 acres of accessible coal, the bed varies from 2 to 5 feet. "In the valley of Mill Creek, it thins out to 8 inches, and increases to 4 feet; if well stratified, it would be of incalculable value; but as it is, mining must always be precarious." [Some of these Ohio "pocket" beds, however, have yielded very handsome profits.] "In S. W. Brookfield, the curvatures are even greater than on the Mahoning; this is the cause of a continual change in thickness, which is liable to disappoint the expectations of the miner at any moment. If the local depression be large, the central part has the full amount of coal, thinning out in all directions towards the edges." In these words, we have a true picture of the way the vegetation was deposited in these earlier beds, upon an uneven bottom around the borders of the coal field, and probably all over it. In this region, the lowermost coals are reported more valuable, and the upper ones (of the few remaining on the ground) valueless, whereas, the upper limestones are numerous and good, thus: "Limestone exists in the greatest abundance in S. Trumbull; the section at Poland exhibits 3 beds in 130 feet; the uppermost 20 feet, has its main development in Pennsylvania, only crowning the heights west of the Mahoning; a hundred feet lower is a hard, blue, brittle limerock, 2 feet thick, which can be polished; and 20—30 feet lower, the same; the second bed, and the lower surface of the third, are fossiliferous. On lot 53, Poland is a calc. shale, 6 feet thick, with myriads of fossils, &c. &c.

30. The following map shows the Ohio descending its central basin; and the Scioto is seen borne off southward by the outcrop of the conglomerate, while the drainage of the conglomerate upper surface is eastward into the bosom of the basin. Whatever irregularities occur are due to the interference of local changes of dip, such as that of 7° N. W. in the Meigs

T. (Muskingum Co.), and in Beaver Co., Pa., below New Castle.

The western edge of the lowest bed of coal crosses Ohio through the following counties, lettered on the accompanying map, beginning at the northeast, Trumbull, Portage, Stark, Wayne, Holmes, Coshocton, Muskingum, Licking, Perry, Fairfield, Hocking, Jackson, and Scioto. Although a general dip southeastward of 10 or 20 feet to the mile prevails between this waving outcrop and the Ohio River, there are

Fig. 23.

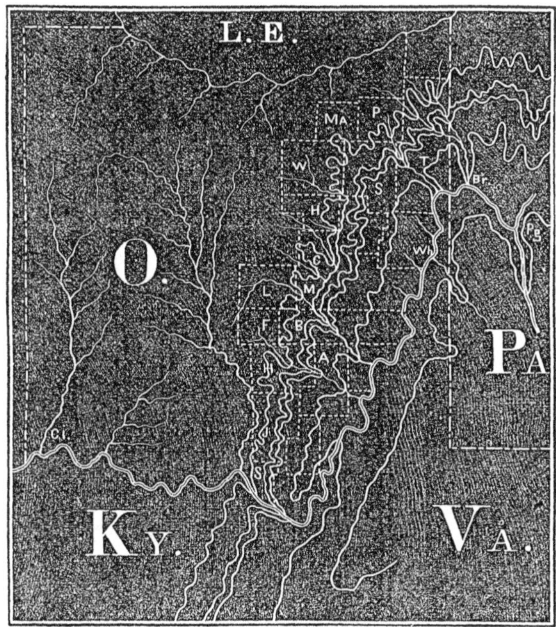

local disturbances of no great length or breadth, which reverse the dip, and form small subordinate troughs, interfering with the identification of the smaller beds and affecting the topography. Three striking facts are visible in this wood-cut: first, the Ohio River runs down the middle of the great coal-field against the shales of the Barren Measures as far as Burlington, and then turns and breaks across the lower rocks towards Cincinnati; secondly, the Scioto flows down against the outcrop of the conglomerate into the Ohio; and thirdly, all the rivers to the east of it, confluents of the Ohio, enter the coal through cross-cut ravines in the conglomerate, and flow down the dip southeastward to the centre of the great basin. In the next chapter, this will be referred to again, as unimpeachable evidence that it was impossible for the rivers themselves to originate this direction, and, therefore, that their valleys were the previous work of other and vaster agencies.

---

31. **The System of the Upper Coals.**—Enough has now been given in illustration of the Lower Series of Coal Measures. The scheme of the Upper Measures, based upon the Pittsburg coal, will now be given from the annual reports of Pennsylvania, and chiefly from the examinations made of it by Mr. McKinley and Dr. Jackson upon the Monongahela, and in Wayne and Washington Counties, where one or two thousand feet of final Barren Measures are piled upon it, and at Wheeling, on the Ohio, where it has been examined and described by Dr. Hildreth, Mr. Briggs, of the Virginia survey, and Mr. Townsend, of Wheeling. Nothing, it must be confessed, could well be more meagre than the information we possess about the last and highest rocks piled in the centre of the great Alleghany coal field; it is a comfort, however, to know that they are comparatively barren of those coal and limestone treasures which it is one principal end and aim of science to place at the ready disposal of society.

Beginning at the top as before, we have at Waynesburg, Green County, Pennsylvania, first:—

Sandstone, 20 ; shale brown and blue, 4, 24 ft.
Coal, 1 foot.
Shale, 10 ; *Sandstone* slaty, 20 ; *Shale*, blue and yellow, 10, 40 "
Limestone, 4; *shale*, soft blue, 4; **Limestone**, 4; *shale*, blue, ferrug. 3, 15 "
Coal, 18 inches.
Shale, blue, friable, 7 "
Sandstone, coarse, brown; large grains uncemented, crumbling into coarse sand; also finer gray building stone, sometimes soft slaty; Brownsville, highest hills; from Greensburg east to river, &c.; graduates downward into shale; fragments cover the hills into Virginia; pounded fragments of it seen near Hillsborough (Wash. Co.), and in Menallan and Redstone (Fayette), 35 "
Shale, brown, yellow, from 2 to 20 feet, 10 "
Coal, good, 6 "
It is wrought in Greene and Washington; central band of soft shale; lower bench best; upper bench slaty; slate pyritous; seen near Brownsville on high hills; 2 miles east of Waynesburg, in the bed of Ten Mile Creek; re-appears on south fork of Wheeling Creek, near State line.)
Shale (sometimes wanting), soft yellow, full of calc. concretions, 5 "
Sandstone and Shale, variable; in Greene Co. solid, light gray, coarse grained; elsewhere thin bedded, soft, shaly. It contains a —— **Coal**, —— 3 —— at Brownsville, Limetown, Elizabethtown, and N. Washington Co., unless this be the bed of coal below, 35 "
Limestone, 8 "
Uniform, five or six layers of hard blue lime, parted by shale full of calc. nod., and in one case a bed of **Coal**, 1 "
Shale, brown yellow, vegetable impressions, sometimes merged in *Sandstone*, flaggy, slaty, building stone (Brownsville), full of mica in sheets, fine grained, light gray; elsewhere non-micaceous compact sandstone, 15 "
Shale, thin, variable, blue, yellow; sometimes contains **Coal**, 1 "
Limestone—Great Limestone, 70 "
This great limestone deposit offers its broad outcrops in all the country south of Pittsburg between the Chestnut Ridge and the Ohio River. It caps the highest hills around

the borders of this area, and forms the beds of a thousand streams within it. It contains no fossils, or the fewest possible, if any. It consists of numerous solid layers, separated by shales from an inch to ten feet thick, which are sometimes true marls; or by calcareous sandstone layers. It is blue, hard, and semiconchoidal; sometimes however a beautiful yellow, and sometimes black, slaty, and fetid, passing into black calcareous slate; passing again southwardly into at least **two coal beds**, one of which has 2½ ft. of good, hard lamellar coal, the other 1 foot. [Three sections made at different places (Germantown, Jefferson, and Brownsville), will show the whole formation to be divided into two by a sand-rock more or less developed, thus: —

Upper member constant,	Limestone,	18,	12,	25
	Shale,	15,	4,	5
Thin gray sand layer			1,	
Limestone and calc. shale,		5,		7
Coal and blackslate,		3.5		
Limestone, hard blue,		5,		
Middle member,	Sandstone,	10,	(calc.) 7,	(slaty) 18
	Slate, black,	3,		8
Lower member,	Limestone,	6,	20,	16

It is probably No. XXIV. of Dr. Hildreth's section at Wheeling.

Shale, yellow, brown; sandy layers; tough hard slate; and fills the cavities of the underlying sandstone with ferruginous clay, shale and impure coal (comp. Hall's *Western Observations.*) 20 ft.

Sandstone, gray, slaty, variable; white cliffs; quarried at Williamsport and elsewhere southward; contains nod. iron and fossil plants carbonized; seems sometimes to rest upon the coal, 25 "

Shale, brown, friable, compact, ferruginous, sandy, with sandstone layers increasing as we ascend; very micaceous towards the north, crumbling and flaggy; sometimes full of plants; lower part often splintery with pencil fracture; surface covered with efflorescence; contains in Fayette Co. a blackslate and **Coal**, *one foot thick;* other thin coal seams run irregularly through it; at Connelsville (Fayette), it contains two beds of nod. ore a few inches thick, 30 "

Pittsburg Coal.

32. The section at Wheeling is as follows:—

Sandstones and **Limestone,**	100 ft.
Coal, 3 feet.	
Shale and Lime,	60 "
Sandstone,	25 "
Shale and **Limestone,**	20 "
Coal, 2.5 feet; fireclay 3.	
Shale and **Limestone,**	30 "
Limestone, 2.	
Shales, &c.,	20 "
Lime (argil.), 15, sandstone 2,	17 "
Lime (argil.), 24; shale 3,	27 "
Lime, 30; broccia 1,	31 "
Lime, 25,	25 "
Slate (micaceous),	9 "
Coal, 1.	
Lime (blue), 13·	
Sand (micaceous),	8 "
Bit. Shale, 5; **Coal,** 1. (ferns),	6 "
Lime (magnes.), 13.	
Lime, 40; fireclay, 4,	57 "
Coal, 1; shale 1; **Coal,** 5.5; fireclay 3,	10 "
Shale, with nodules of limestone, 5; blue, 10,	15 "
Sandstone, quarried,	25 "
Shale and Sands, clayey, mic. calc. ferns,	17 "
Lime, hard gray, few foss.,	
Red *Shales,* calc. variegated; olive and green below,	70 "
Ohio River.	

Fig. 24.

Fig. 25.

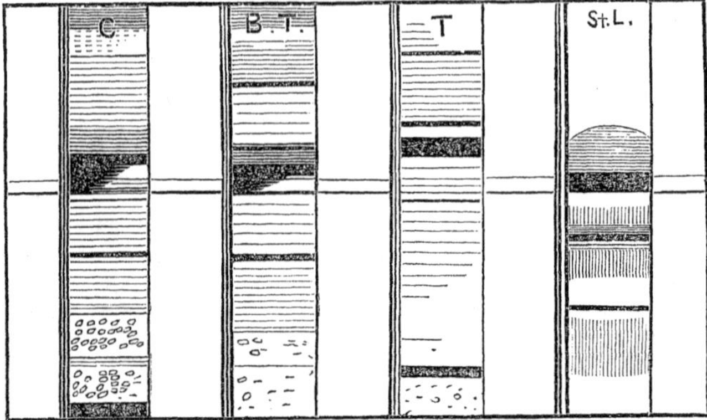

Bottom of the Coal Measures, p. 93.

C. Cumberland Mountain, Sewanee Mines, Tennessee.
B. T. Broad Top, Riddlesburg bed, Pennsylvania.
T. Towanda Mountain, North Pennsylvania.
St. L. St. Clair Mines, Illinois, opposite St. Louis.

TOPOGRAPHY.

CHAPTER III.

TOPOGRAPHY—AS A SCIENCE.

1. **The Science of Topography,** like every other science, proceeds to deduce from a few elementary laws an infinity of forms, by which these few laws may or actually do express themselves upon the surface of the earth. The possible is always an infinite series, the actual a limited and fortuitous selection from it—not always, therefore, the most striking, perfect and complete. Here and there, now and then, one such occurs and we call it a typical phenomenon, because, like the Apollo or the Venus in human Fine Art, it fills out the expression of those laws or creative ideas, the elucidation of which alone is science and the standard of value for a fact.

2. In the coal regions of America from which most of the illustrations in this book have been obtained, one group of topographical expressions are in great perfection, but others are entirely wanting. For alpine forms, that is, for the aiguille or needle peak—for the jagged crest line—for the cirque of nevé wherein glaciers form—for roches moutonnés or polished bossy slopes where glaciers have passed—for moraines or piles of fragments left by glaciers in their retreat—for viæ malæ or profound gorges with fitting walls and dark recesses cleft by earthquake action, and too high above the later sediments to be filled up—for perpendicular horizon outlines, thousands of

feet long against the sky—for thread-like waterfalls pendant from the edges of pastures, hundreds of feet above their base —for rosaries of lakes filling with mud and sand, protecting others below them, as clear as crystal, and lined at the bottom with the thinnest strata of impalpable powder, and of filled up lakes, now fertile meadows, supporting secluded hamlets—for such phenomena as these, we must go to the Alps, to the Pyrenees, to the Andes, to the Himalayas.

3. In the coal measures we have, on the contrary, long and regular mountain crests vanishing horizontally in parallel lines into the horizon, broken at intervals with gaps and curving in and out in zigzags including inner sets of zigzags, system within system, down through the series of the rocks. We have all the phenomena of drift and denudation, deluvial scratches, eddy hills and terraces, in great perfection.

4. In that Appalachian region, as it is now called, which, as a belt from fifty to one hundred miles in width, between the edge of the Bituminous Coal region and the South Mountain or Blue Ridge, stretches from the Delaware to the head waters of the Tennessee, we recognize a country where, from the comparative deficiency of fossils and of coal, and from the repeated and elongated outcrops of a few valuable mineral deposits, the science of geology transforms itself into the science of topography. Nowhere else on the known earth is its counterpart for the richness and definiteness of geographical detail. It is the very home of the picturesque in science as in scenery. Its landscapes on the Susquehanna, on the Juniata and Potomac are unrivalled of their kind in the world. Equally beautiful to the eye of the artist is a faithful representation of their symmetrical, compound, and complicated curves upon a map. The region of the European Jura resembles it in general structure, and exhibits the ancient action of similar forces under the same laws, but in less detail, and with far less delicacy. The author was fortunate in being the first geologist who had an opportunity to approach the dynamic phenomena of the Jura with an American eye trained on this

typical ground. This was in 1844 ; Mr. Rogers did the same in 1849. M. Desor, the distinguished collaborator with Agassiz, of the Bernese Oberland, and now filling a chair at Neuchâtel, will also testify that here alone, in the Appalachians, he could obtain those constructive ideas which were needful there to comprehend the puzzling arrangements of his native mountains. Years of patient toil it cost us to unfold the mysteries of the Pennsylvanian and Virginian range—a tangled hank, to be untangled thread by thread and rearranged skein by skein—a tracery more elaborate and intricate than Gothic or Arabesque—nature's primeval labyrinth, in which the minotaur was but a form of science cast in metal and sculptured in stone; a sphynx whose riddle has at length been read and written out by men like Henderson and Whelpley in what are now to be forever the hieroglyphics of geology. Their manuscript maps, executed under singular difficulties, now nearly fifteen years ago, have disappeared, but can never be forgotten.

5. That the European Jura, although the home of geology for at least a century, had to wait for its elucidation until the American Appalachians had been mapped may seem strange, but it admits of easy explanation. In Pennsylvania palæontology and detailed local stratigraphy were impossible ; we were untrained in both. We had an empire to survey and little time to do it in. The country was no ground for mineralogy; no rare and curious minerals exist in it; it is a waste of sand, mud, iron, and limestone strata of various texture, and color, in endless repetition; to know one was to know all, to know it here was to know it everywhere. Nothing remained to study but dynamic forms ; and these so numerous, so grand, so variously grouped that they excited a perpetual enthusiasm, and led on to infinite research ; they supplied the place of fossil forms, of forms of crystals, and of variations of mineral elements ; they were a world of the exhibition of natural forces by itself, and as such we took possession of it and settled in it as our fathers did in the valleys themselves, and thus became

not mineralogists, not miners, not learned in fossils, not geologists in the full sense of the term, but topographers, and topography became a science and was returned to Europe and presented to geology there as an American invention. The passion with which we all studied it is inconceivable, the details into which it led us were infinite. Every township was a new monograph. They trained us to that fertility of fancy which precedes the ripening of the *constructive* or geometric faculty, made to act upon the more difficult problems of geology; while its generalizations were so vast that to our eyes, familiar with rock curves ten and twenty miles in radius, underground or in the air, all that customary European local research which filled the proceedings of societies in lieu of new and larger matter seemed tedious and puerile. The moment, therefore, one of us beheld the ranges of the Jura, with their combes and offsets, their vast escarpments, and far glittering white gaps, he felt at home among friends where geologists born among them felt that they were strangers. For the valleys of the Jura are filled with later formations full of fossils, which Appalachian valleys never are. There much is hidden, here all is told. The fossils themselves in the Jura distracted study from the topography. The topography also is gross, massive, simple, destitute of those lessons in detail, those innumerable repetitions in petto of the grander curves, from which we got our first instruction and to which we retreated from insurmountable difficulties to learn how to return and overcome them. The small neat echelons of Mercer and Juniata were preliminary problems to the vast and forest-covered zigzag steeps of Union and Schuylkill Counties. So are the latter to the yet more titanic groups of lines and curves of Bienne and Neuchâtel, which front the still sublimer and still uncomprehended Alps. The influence of the local genius was exerted with equal energy upon the New York geologists. The long thin horizontal outcrops of the Pennsylvanian system in New York, without dynamic features excepting a few cross faults, in the valleys of the Hudson and

Mohawk, led Vanuxem, Conrad, and James Hall into the almost exclusive discussion of minerals and fossils, and made the last our great American palæontologist, the master in whose words we all now swear; while Troost and Owen, and the rest obeyed a similar local and irresistible commandment of nature in the Western States. But the coal with its Vespertine foundation was wanting to New York; it remained therefore for Lesquerox to inaugurate the science of its flora.

6. We cannot conceal from ourselves the significance of the jurassic representation of Appalachian topography. It is no mere fortuitous resemblance; it is not even an accidental repetition of the phenomena of volcanism at the same moment at two distant points; it has a much deeper meaning in cosmology; it shows the permanence of geological dynamics, the continuance through the world's history of certain great laws of mechanical force. For the Jura are young enough to be the grandchildren of the Appalachians. Long after the strata here were thrown into their waves, and the surface had been cut out accordingly into its lines and curves, and eastern North America had become a veritable *continuance*—a continent—a fixed fact, the deposits of Europe were still forming around the islands of its archipelago, the Cantal, the Taunus, the Thuringinwald. Ages of creations followed before the then antique topography of the "New World" was suddenly and grandly imitated in the old, and the waves of the Jura arose obedient to the selfsame formulæ, and were carved into similar lines and curves. Then Switzerland, France, and Germany became likwise a continent, not by invention so to speak, but by tradition. Whether the oriental mountains, the ranges of the Koordish New Red sandstone, that support the plateau of Persia, are or are not a third and equally noble instance of Appalachian dynamics and topography no scientific traveller has yet announced, and none but an American geologist could at present demonstrate.

7. All topography resolves itself into a discussion of the

Mountain. The surface of the earth may be considered as a congeries of mountains (and hills are but smaller mountains) touching each other at their bases. Valleys have only a negative topographical existence and represent the absence or removal of mountain land. Mountains are solid portions of the earth's crust, while valleys are but vacua in it, taken possession of by the water and the air. The propriety of this view, opposed though it be to the economical history of the human race, will become more evident when the denudation of valleys is discussed. In geology it is important to true views to consider all changes to be mountainous, and valleys to be an incidental consequence.

8. **A Mountain** has **three Elements**, top, side, and end, and the primary discussion of a mountain is that of its slopes, its crest-line, and its termini. Gaps are irregularities in its crest line, as terraces are in its slopes. The cross-section of a mountain shows why its slopes, and usually also why its crest-line and its termini are not only what they are, but could not be different. It is a deep set feeling among men that if there be accidental forms upon earth they are to be found in mountains. There could not be a greater mistake; for if there be natural forms unalterably predestined by the direction and intensity of natural forces they are those of mountains. Not a wrinkle in the side, not a notch in the crest, not a flexure in the trend of a mountain or a hill but is an evidence of laws which have operated upon it with the nicest precision. Not a ravine, not a rod of cliff, not a waterfall, but exists in the immediate vicinity of its own explanations. The place where a stream breaks through, however apparently accidental, was determined by positive relationships between the rocks of the locality; nor can any investigation be more exciting than that which rewards itself with perpetual discoveries of cause and effect in a wilderness of apparent lawlessness and unexplained confusion.

9. **The Mountain Slope.**—To illustrate the influence

which its interior structure has upon the form of a mountain, the accompanying series of cross sections are introduced. The first set show how a stratum of sand-rock or other hard material inclosed in softer stuff, arranges the height and slopes

Fig. 26.

of its mountain to suit its own dip. When it is vertical the mountain is low, sharp, and symmetrical; at 60°, it is higher with a front side long; at 30° higher still, with a long, gently sloping back, and short steep front covered with angular fragments from a range of cliffs above; when horizontal, the mountain is at its maximum height, forming a table land with precipices and steep slopes in front. It is needless to suggest the infinite variations of this simplest law of mountain form.

When two such sand-rocks lie in neighborhood they of course form a double mountain, subject to the same vicissitudes of external aspect in view of similar changes of internal structure. It is not so easy to show the effect of these changes under another influence, that of subsidence beneath a universal plane of denudation. The attempt, however, is made in the following sets of cross sections. It must be understood that a mountain whose rocks stand vertical or horizontal at one point may show them much inclined at others (its form will change to suit), and also, that as mountains are but fragments of the upper layers of the earth's crust preserved from the general denudation and translation by lying lower than the rest, in hollows or synclinals, as they are technically called, these synclinals as they rise above or sink below the average level or line of denudation will give up to destruction more or less of their contents. In other words, when a geologist traverses one of these geological basins lengthwise he finds the highest rocks in its deepest parts, and the lowest rocks in its shallowest or highest parts. The cross sections given below are arranged to show the coming in of higher and higher rocks as the basin sinks, and the consequent changes of form which the mountain or mountains undergo.

10. In every **Shallow synclinal** there is a high, narrow, flat-topped mountain, with precipices on both sides looking down upon the lowlands. Such is the structure of the Catskill, Towanda, Blossburg, &c., and other mountains in the north, and the Cumberland Mountains of Tennessee and Alabama. As the basins deepen, other higher sand-rocks come in above, and double the precipices and slopes; finally the whole is cleft in two, and the drainage, after traversing the parted mountain often for many miles, breaks out sideways into the plain. It is evident that it was due to the direction of the original currents setting along the centre of the geological basin before the cutting developed the present mountain.

Fig. 27.

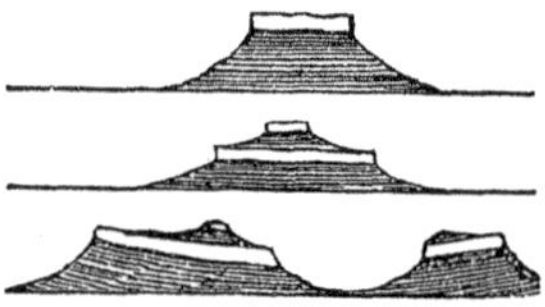

11. The **Sharp synclinal** shows this more clearly, with this difference, however, that its longitudinal central cutting is never a ravine, but always a valley. The hard rocks pass off in diverging crests terraced and gashed, inclosing one another, and giving place to mountain within mountain, in a series that has no limit but the number of hard beds in the system of formations. Their precise inclination with the horizon makes no essential difference in the action so long as the synclinal remains a simple one; but the moment this becomes compound then all manner of complications inaugurate themselves upon

Fig. 28.

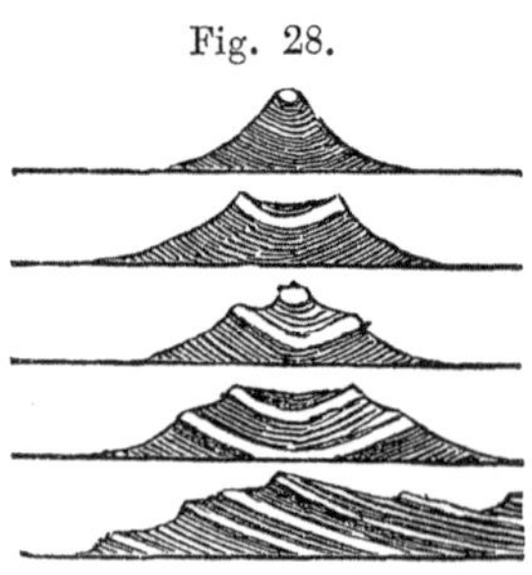

the surface and present a thousand puzzles to the skill of the geologist, as in the following set (Fig. 30), which more or less nearly represent the wrinkled compound synclinals of the anthracite coal region.

Fig. 29.

12. The **Anticlinal** structure is the reverse of the synclinal, and has its own infinite system of forms equally subordinated to the general laws of denudation, and in many respects curious inverted parodies of those above. In fact, as may be seen in the next set (Fig. 31), the moment the huge back of an anticlinal mountain splits into two, we lose the anticlinal as a character of the mountain while it remains in the valley, until from the centre rises another mountain, to be again split lengthwise in its turn. In this case we represent the anticlinal as rising slowly, just as before we represented the synclinal as slowly sinking. The ridges on each side of a split anticlinal are called *monoclinal*, and are in fact the same as the ridges into which a parted synclinal mountain divides itself. Hence the forms are common to both.

Fig. 30.

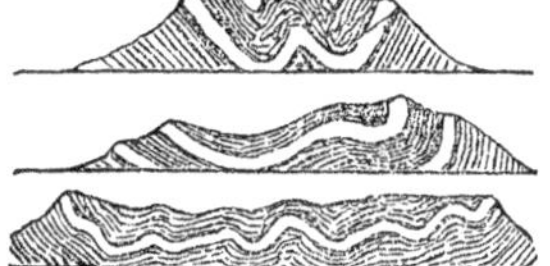

Fig. 31.

13. Much is here committed to the genius of the reader to study out and comprehend. There are certainly great obstacles in the way of a clear conception of these laws of force and form, but it can be got, though it can hardly be given even by the utmost "popularization" of scientific terms. This manual

is intended to assist only the earnest student, and the illustrations adduced are sufficient, if well pondered, to reveal the whole mystery of mountain cross form or slope so far as the sedimentary rocks are concerned, and these alone include the coal.

14. The **apparent slope** of a mountain is always exaggerated by the eye looking at it from above or below, and therefore vastly foreshortened. Travellers make the absurdest blunders on this point, talk of looking down *precipices* which would not measure more than 30° or 40°, while a true preci-

Fig. 32.

Val de Drac, Dauphiné.

pice is 80° or 90°. Nothing but experience, and the honest use of a slope instrument will cure this prejudice. It should be recollected that dry sand or gravel begins to move down a slope when it reaches 30° or 35°; and a man must scramble on his hands and knees and cling to twigs and stones upon a slope of 50°. We have such slopes almost nowhere, except on the metamorphic mountains of Tennessee. Most of our steepest mountain slopes do not exceed 40°; this is not true of other countries. In (Figs. 32, 33) sketches of slopes in the French

Alps above Grenoble, the greatest care was taken to draw them truly on the spot, and reproduce them truly on the wood. The slopes are too distant to be interpreted geologically, but it is very evident that they bear no affinity to those of **unchanged** sedimentary strata, like our own.

Fig. 33.

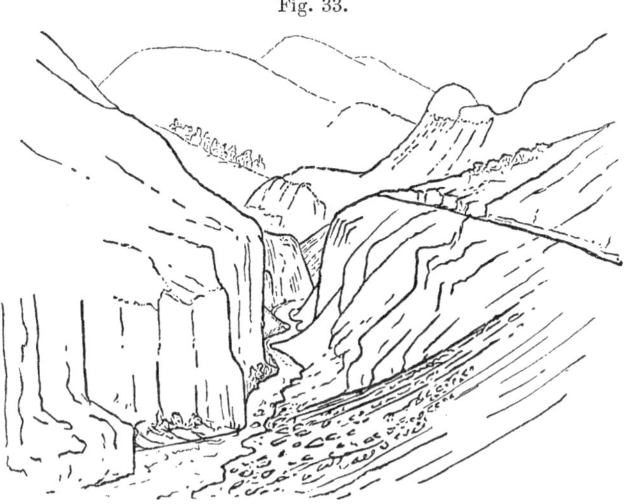

Val d'Oysans, Dauphiné.

An explanation, however, may be found of these immense walls (true precipices, even more than vertical), by regarding another sketch (Fig. 34), made with equal care, in the Hartzgebirge of Germany. It shows the chasm-valley of the Bode, as it appears on looking down from the Ross Trappe, or Hoof-Print Pinnacle. The whole mass of mountain is of an altered ancient slate, vertically fissured and laminated, and therefore denuded in walls. Similar appearances occur in the Pyrenees, but among quite different rocks. Aiguilles are a variant of this form, but should be discussed with crest lines.

15. **Volcanic slopes** are of a different order from those we have been considering. Generated in air by gravity, they cannot conform to the laws of denudation. Composed of fine

dust, pumice-stone, sand, and stones ejected from the crater when red-hot, and falling in parabolic curves upon its sides, they form a porous, bell shaped cone, which absorbs instantly the most destructive torrents of rain, yields no springs to wear away the sides, and covers its fertile soil with a still further

Fig. 34.

Der Bodenthal.

protection of furze. There seems no limit to the possible age of the permanent form of these cones, after the underground fires have gone out. In the case of living volcanoes, like Etna, the flow of lava down their sides produces an extraordinary complicity of outline, in the general form of a very flat irregular cone, studded with smaller cones, with radiating rents descending from the summit.

16. **Trap hills,** however, and other ancient lava outcrops, so frequently seen among the older rocks, obey the same laws of denudation with sandstones and marbles. For the present, we take for granted the fact of an immense destruction of the ancient surface, the result of which was the construction of the present topography, which can be accounted for, in fact, on no other hypothesis. It is acknowledged, on all hands, that the only question open to discussion is, whether this planing down

of the crust to its present surface was a secular or an instantaneous work. Many of the ancient, like some of the more modern lavas, were ejected under water, or at least flowed down under water, and spread themselves over the bed of the

Fig. 35.

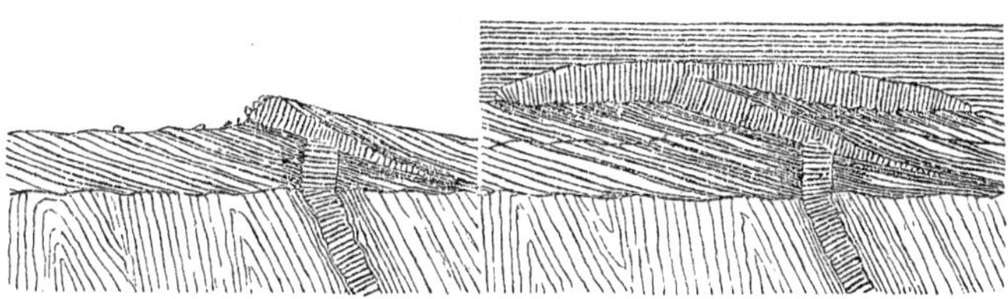

sea. The accompanying cut of one of the many trap ridges of the New Red Sandstone region in Pennsylvania, New Jersey, and Connecticut, represents the original sea, the lava and its vent, the manner in which it lifted the new red layers at its outburst so soon as it was near enough the surface to do so, and overflowed them above; and opposite, the present surface with all the materials above abraded and gone and the steep columnar cliffs of greenstone trap covered with fragments. The red sandstone on the back of the ridge is seen on examination to be altered in color and texture by the heat of the trap so that its increased hardness assisted in the formation and determined, in part the shape of the hill.

17. All **Injected rocks,** such as the granites, syenites, trap, basalt, &c., as a rule, form hills, partly by virtue of their own crystalline hardness and massiveness, but chiefly by semicrystallizing and hardening the rocks through which they pass. If these be homogeneous clays free from mica, such as would form valleys in their unaltered state, they become excessively hard, and tower into the air as mountains. If, on the contrary, the heat develop talc and mica shists, what would have been a mountain land sinks into almost a plain.

18. **The Crest Line** of a mountain is normally either a point or a horizontal line, according to whether it is a mountain of ejection under air, or a mountain of elevation and denudation under water. The type of the volcano is a cone; the type of any other mountain is a double sloping wall. Sedimentary mountains are the turned up edges of some hard sediment. The central hardest or massivest layer of that sediment will of course form its crest. If there be subordinate layers, they will form terraces; and, at the gaps, ribs. If several central layers near together be equally able to resist denudation the crest will move from one to the other along the line according to the cross-cutting. So long as the dip maintains itself steadily the mountain crest will be a straight and level line; when the dip changes, the crest line will fall off accordingly and suffer change in height and evenness. When the dip falls completely over or reverses itself, the crest line will double back upon itself, and do so as many times as the dip changes. At such a recurve coves are formed; here valleys head up; here gaps usually break through the side walls; here also the mountain attains its maximum of height. These doublings, if anticlinal, are in one direction; if synclinal, in the other. A series of them can only occur where a system of parallel anticlinal waves and synclinal basins pass under an outcrop, as in the following fine instance.

It represents the Buffalo Mountains, L of Union County, Pennsylvania; Jack's Mountain, K; the anticlinal Kishicoquillas Valley (Logan the Indian lived at G); the upper end of the synclinal Standing Stone Valley H, and the Seven Mountains back of it; the Bald Eagle Mountain sweeping from Bellefonte J, to Muncy D, on the Susquehanna West Branch; the Nittany Valley A, with its anticlinal limestones appearing again in the Nippenose or Oval Valley B and the smaller Oval valley C; the Nittany synclinal mountains south of it with their small included anticlinal vales; and finally the simple anticlinal Montour's Ridge E apart from all the rest.

This magnificent group of **Zigzags** (studied out by Mr. Alex.

Fig. 36.

McKinley) has probably no rival yet known upon the surface of the earth. Until its structure was explained it was a confused wilderness of mountains covered with forest and interpenetrated with a network of ravines; some of its crests long horizontal lines of singular evenness, and others broken by a series of sharp notches or deep gaps. Its most singular feature is the terrace structure of its valleys running into a **reversed ship's keel structure** in the mountains which inclose them; a structure which needs no description when the reader turns from a perusal of the map to the third cross-section in Figs. 28 and 29, pp. 128, 129. These mountains are all composed of the Second Great Sand-rock described in the last chapter, the lower Silurian, Shawangunk Grit, No. IV. The terraces are made by the lower of its two hard members, the keels and crests by the upper. The included valleys are all of No. II Trenton Limestone, with No. III or Hudson River slates around their edges. The outside country from Lock Haven R, round to Sunbury F, consists of Upper Silurian through which the Lower Silurian projects in the broad-backed anticlinal Montour's Mountain, E.

This map, rough as it is, will reward an attentive examination. The notching of the *southern* leg of each anticlinal zigzag by the rush of denudation across it, from within outward, or *vice versa*, is particularly interesting. The peculiar drainage of the head of the anticlinal, not inwards towards the central valleys, but sideways through into the outside synclinal coves is also very instructive, and speaks volumes for the elucidation of the denuding action.

19. All the phenomena which, in the preceding wood-cut are confined to the mountains of IV, from eight to twelve hundred feet high, occur in more delicate detail in the less elevated ridges of the softer upper silurian rocks of VI, VII, and VIII, in the region of the Juniata, studied by Dr. Henderson, from a portion of whose MSS. topography, the author has been kindly permitted to make the following map; and,

Fig. 37.

Ch.
L.Ak.V.
A.V.
T.C.V.
B.E.V.
H.V.
N.H.
M.V.
Car.
T.V.
S.R.
C.
D
B
O.R.SS XI IX
O.SS. VIII VII
Cl. V
H.R.S. III

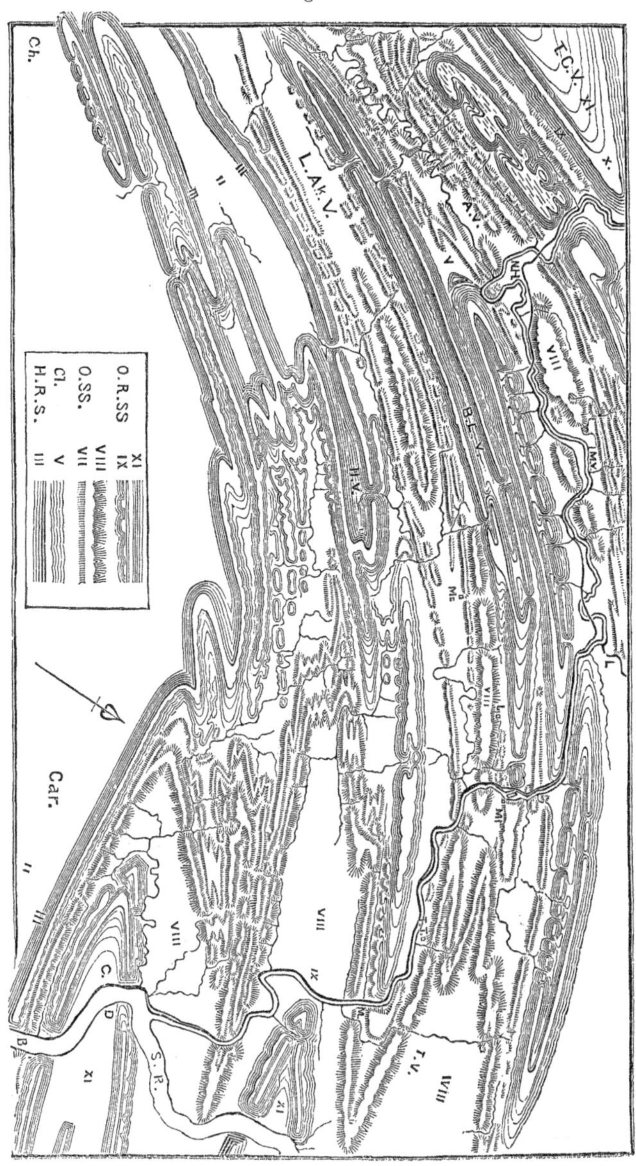

in presenting it, would take the responsibility of saying that, certainly, up to that period, if not up to the present date, no piece of geological surface painting in any part of the world of equal finish and expression, has appeared. It is a *chef d'œuvre* of geological acumen and constructive genius in field work, impossible to excel, and which, under any other organization of the Pennsylvania State Survey, would have made the name of him to whom we owe it, fifteen years ago, familiar to every scientific ear in Europe.

Description.—The letters on the map read as follows :—

T. C. V. Trough Creek Valley.

N. H. Newton Hamilton,	J.	A. V. Awkwick Valley.
M. V. McVeytown,	M. E.	L. Ak. V. Little " "
L. Lewistown,	} Upon the Juniata.	H. V. Horse Valley.
Mf. Mifflintown,	Ch. Chambersburg.	B. L. V. Black Log "
T. V. Thompsontown.	Car. Carlisle.	T. V.

C. the Cove; D. Dauphin; B. The Penna. R. R. Bridge over the Susquehanna.

In the square, the letters read O. R. Old Red Sandstone; O. S. S. Oriskany Sandstone; Cl. Clinton Group; H. R. S. Hudson River Slates. The Roman numerals scattered about the map, show the formations according to the Pennsylvania enumeration, the equivalents of which may be found in Appendix VII.

20. The **geometrical construction** of these topographical **zigzags** is perfectly simple. When a cylinder is sunk obliquely in water, the edge of the water describes an ellipse; and if several be so immersed side by side and touching each other, the water line will be a zigzag, Fig. 38. By varying the shape of the cylinders, by making them flat or sharp, or by using prisms and cones of different orders, an infinite variety

Fig. 38.

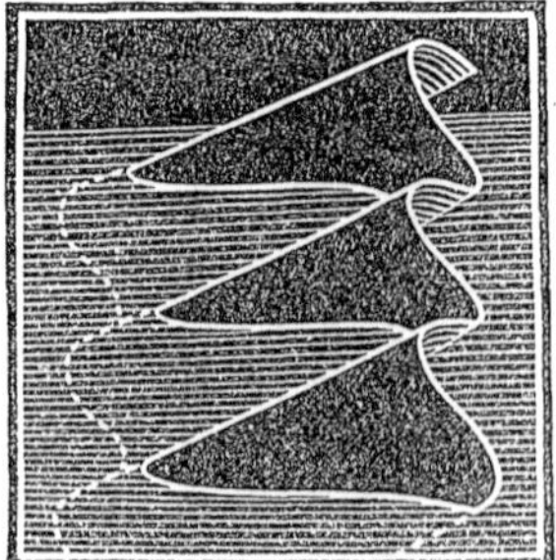

of zigzags will occur. If now the anticlinal waves be taken as so many cylinders or cones in lateral contact, and the general horizon of the country as an approximate water line, which it is, all the geometrical conditions of the topography are made out. Every ellipse on the map points in one direction, over a declining anticlinal underground, and along an ascending synclinal in the other. If two anticlinals decline side by side in opposite directions, they describe upon the map, if the included synclinal be deep enough, two separate ellipses pointing opposite ways (*a*, Fig. 39); if the synclinal be too shallow, the ellipses are resolved into two zigzags in reversed relation to each other (*b*). If the anticlinals increase in height side by side, the ellipses lengthen out side by side in echelon (*c*). If the anticlinals are of equal lengths and heights, dying both ways, but reach their maxima in advance of one another, there result two correlative sets of zigzags both in echelon (*d*). If as is commonly the case, a middle anticlinal of great height has on its flanks a diminishing series of anticlinals, the zigzags fall into a rhombic train (*e*). If two such systems of anticlinals coming from opposite regions where each is at home, pass each other, both being in decline, the structure of the Anthra-

Fig. 39.

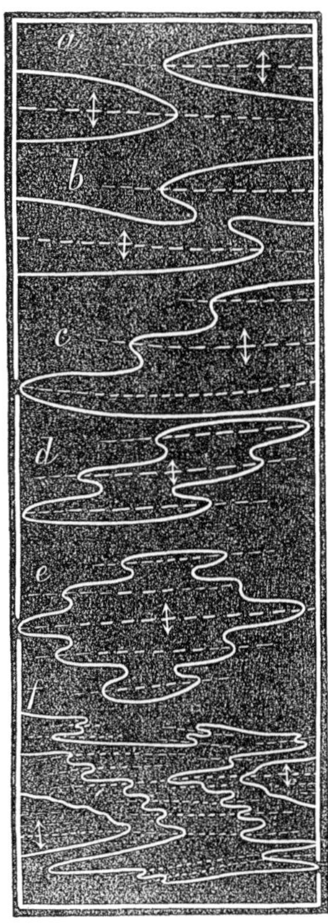

cite Broad Mountain (*f*) is the consequence, *g* being the great compound anticlinal of the Mohontogo Mountains; *h*, the great compound anticlinal of the Pohopoco, Pocono or Nesquehoning Mountain, and *i*, the coal basins on the intervening plateau of the Broad Mountain.*

Fig. 40.

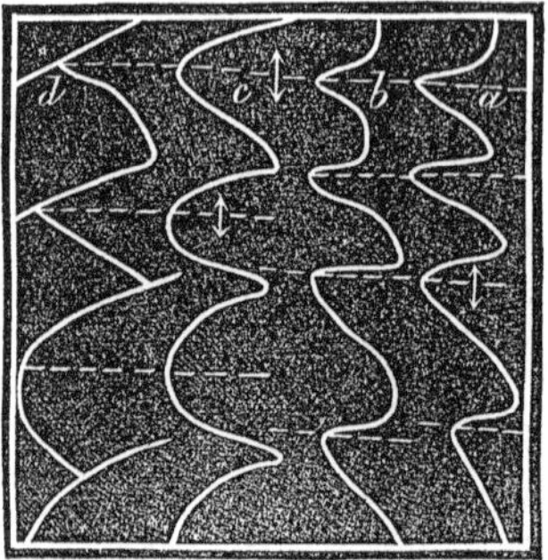

21. So much for position. As for the **shape**, if an anticlinal is steeper on the one side than on the other, the ellipse will be unsymmetrical, like (*a*, Fig. 40); if it be sharper above than the synclinal is below, it will draw the curve (*b*); if the synclinal be sharper than the anticlinal, the curve (*c*); if a break occur in either, the curve will resolve itself into (*d*) or (*e*). It is not necessary to carry the synclinal water lines through its series of forms, for they are involved and in fact identical with the anticlinal series, forming the return curves of the latter.

22. In the *anthracite region*, the same laws are seen at work in the Devonian and Carboniferous Formations, IX, X, XI, XII, and the coal. These were mapped in 1840 with equal care, but under greater difficulties, by James D. Whelpley, and the general structure perfectly made out; his own large map has never been published, but was compiled upon a reduced scale with the others to make the Geological State Map, said to be now in the engraver's hands abroad. The same laws governed a similar topography throughout Virginia until we approach the Tennessee line, where the anticlinals pass into straight fissures, and the zigzags resolve themselves into perfectly parallel straight mountains, abruptly ending.

* Unfortunately, this diagram was not reversed in the cutting and therefore stands reversed upon the page.

One marked peculiarity of all this region is seen to be the synclinal and anticlinal canoes.

23. **Canoe Mountains** are of two kinds, distinguished by their walls, and by their ends; in every particular, they are the reverse of one another; in fact, one is merely the other turned inside out. Three **anticlinal canoes** are seen in echelon at the top of Dr. Henderson's map; they are the denuded backs of three anticlinal waves of Lower Silurian, filled with Hudson River slates III, with a central streak of Trenton Limestone II, *terraced inwardly*, and with *long, pointed, subsiding ends.* **Synclinal canoes**—the ends of two are seen in the same map to the right, composed of Devonian, and truncated by the Susquehanna—are terraced *on the outside*, and have *high, blunt, symmetrical, precipitous ends,* from which, such as the Pocono back of Stroudsburg, the Shickshinny north of Catawissa, the Terrace Mt. south of Huntingdon, &c., the finest views in the region are to be obtained. But by one feature especially the *anticlinal* canoes distinguish their crest line at the termination; by its superior height and side gaps. The folding of such a mass of rock upon itself has here offered the greatest resistance to the denuding agent, and that, too, just where the latter had freedom to pass around; in consequence of this, the crest is lifted into the air more boldly here than anywhere above its line. This is true of every *sharp anticlinal* canoe mountain. A fine example, seen from a distance, is given in the annexed sketch of Peak Mountain, Fig. 41, northeast of Wytheville in Southern Virginia. We are looking at its northwestern side from the southwest. The sketch of Tussey Mountain from the top of Terrace Mountain, back of Stonerstown, in Huntingdon Co., Pa., Fig. 42, is added to show the anatomy of this feature. We look into the canoe over its broken and gaped wall. The hills in front are low ridges of Upper Silurian, dipping towards us away from the anticlinal in the canoe.

24. The crests of these mountains are more or less torn up, and notched or gaped according to their local exposure to the denuding agency; and the facts coming under this head

Fig. 141.

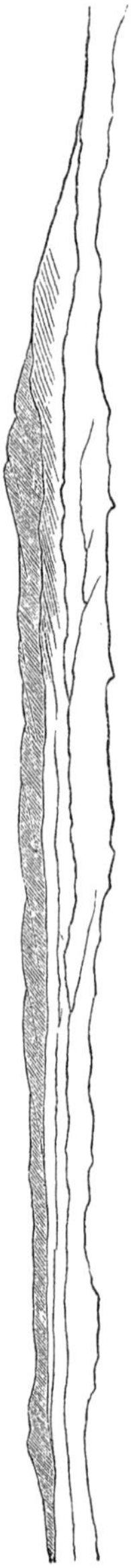

are of themselves sufficient to prove that it could have been no other than a fluid, a vast body of water moving vehemently across the suffering edges of the strata. Where these edges were long, straight lines, the flood seems to have passed over them in a smooth sheet leaving a horizontal, even crest, broken by a single deep gap, or at most by two or three, out at which the final drainage passed; but wherever a system of small anticlinals complicated the outcrops and condensed and deflected the waters, the whole line of crest is broken up by innumerable notches and deep gaps. Where the strata recline at moderate angles, from 30° to 50°, the crest resisted with more success and is left more regular; wherever, as in the Bald Eagle, the strata stand vertical, the crest has suffered immensely. The gaps at the ends of the anticlinals of IV, above described, have the shape seen in this (reversed) sketch of the ends of one of the Buffalo Mountains. Both the gaps seen in the sketch show the normal form proper to all Appalachian gaps, and almost always noticeable, viz: one side steep, sharp, and high above, the other side steep below and sloping low and long above. This unsymmetrical structure is due to the set of the diluvial current, and is always connected with the phenomena of eddy prongs and cones to be described in the next pages.

25. The **roches moutonnes**, or embossed slopes of Alpine valleys, produced by the slow movements of glaciers, and replaced above a fixed line by taluses, or straight slopes of broken materials fallen from vertical ragged cliffs above, towering into still higher peaks and needles of clayslate, are nowhere to be seen in the Appalachian Regions. Not a trace of

glaciers is to be discovered by the most diligent search, no embossed slopes, no moraines. Whatever produced our topography, certainly *glacial ice had no hand in it.* There is drift, but it lies in sheets from 10 to 50 feet deep on *the sides*

Fig. 42.

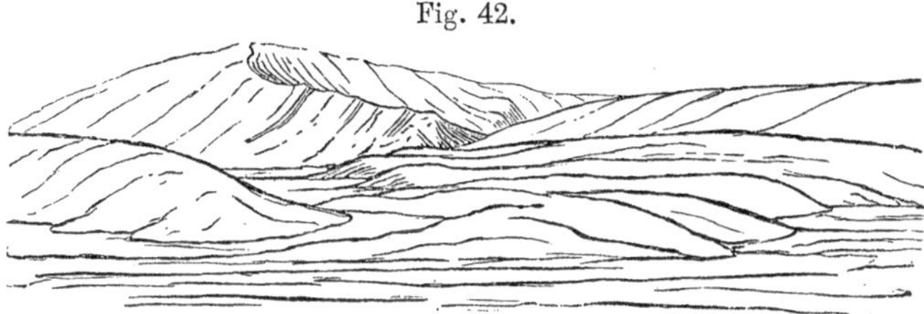

of the gorges descending the Alleghany Mountain, and *at their heads chiefly,*—also behind the gaps in the lower regions, in banks at the upper ends of the valleys, in fact, wherever our knowledge of *retarded floods* behind barriers or obstacles would lead us to look for coarse mechanical deposits. The Jura, on

Fig. 43.

the contrary—although possessing an original topography strictly Appalachian and presenting all the phenomena of synclinal and anticlinal zigzag ridges, canoe valleys, terraces, and gaps—differs from our mountains in this, that subsequently to the finishing of its topography it was invaded by glaciers, and these, although now no longer in existence, have left their usual traces in scratched and polished surfaces, rounded shoulders, limit lines, side moraines, and, above all, semicircular end moraines, or piles of rubbish rock brought by the ice from the interior valleys through gorges, and deposited outside in moon-shaped rows, as in the following copy of one of Prof. Agassiz' lecture illustrations. Nothing of the kind has ever yet been made out in Appalachian topography; nor could the Jura have borne such children but by direct impregnation from the Alps before her. When the Alpine glaciers had swollen to such an extent as to issue upon and cover over to the depth of several thousand feet the great lowland of Switzerland and embank against and invade the Jura, the reflex action of this force so foreign to the Jura, produced from its valleys, on the retreat of the ice, all the phenomena of an original glacial region. It is yet to be demonstrated—and the author has very little faith in its being soon done—that even the White Mountains, semi-alpine as they are, have ever played this game with the Green Mountains, or have ever themselves produced true glaciers. The diluvial grooves and scratches and the forms of the slopes of the Welsh Mountains, seemed also to him, when on the spot, to belong to Appalachian topography exclusively, and not to aerial, glacial, or Alpine surface forms.

Fig. 44.

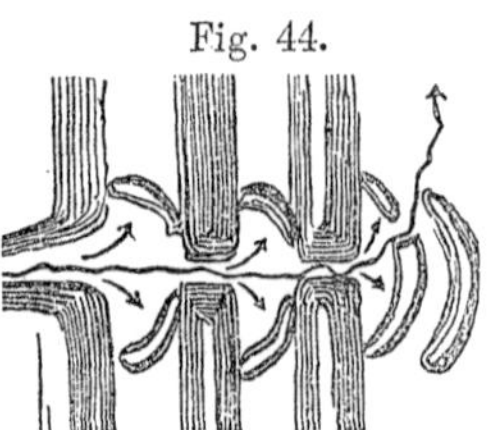

26. The **Terminus** of the mountain remains to be discussed as the third prime element in topography. It is, however, so involved with the discussion of the crest line (of which it forms as much a member, as the summit does of the slope), that little remains to be said of it. The law of the termination of ridges is simply that they end with the rock which makes them. This happens in three ways:—

27. a. **Mountains terminate when the rocks thin out.**—There is no formation that envelops the whole earth and very few that can be traced a thousand miles. The majority of minor sand-rocks, hard limestones, compact slates, outbursts of trap, plates of lava, veins of granite, the outcrops of which form the subordinate ridges of a region, grow thinner and thinner in both directions from a centre point and at last

disappear. Of course the ridges which they make become less prominent, lower, more ragged and uneven, run into lines of isolated hills and low knobs, and finally sink into the plain, or coalesce with the sides of other larger ranges of more persistent rocks. Innumerable instances of this occur; Henderson's map shows several. The hills of For. VII. are characteristically subject to these changes. When the formation thickens or "comes in again," its strengthened outcrop renews an elevated range of hills. But the great sandstone formations described in the last chapter carry their mountain outcrops from the St. Lawrence to the Tombigbee.

28. **b. Mountains appear to terminate in folds.**—All sedimentary rocks of ancient date have been laterally pressed into folds or anticlinals. These have been described already as forming an outcrop in zigzag and echelon, and appear to bring the mountain to a series of terminations (as in Fig. 36) long pointed and sinking slowly into the plain in one direction, and to a corresponding series of high abruptly terminating knobs in the opposite direction. But in reality the mountain simply oscillates between these as angles of direction and then passes on in less disturbed lines as before. The great Kittatinny or Shawangunk, Blue or North Mountain (for by all these names and many more is it known along its course, beginning at Newburg on the Hudson River and running into Virginia across all the rivers which descend from the Alleghany Mountain into the Atlantic Ocean), is the best example we possess of this curious and beautiful phenomenon. The small anticlinals which penetrate it at a low angle with its course, usually entering its southern flank from the east, cause it to fold upon itself five times in New Jersey, twice at the Delaware Water Gap, once at the Wind Gap, ten times at the Little Schuylkill, once at the Swatara, twice near Carlisle, and twice again before it enters Maryland. In smaller ridges this difference between the direction of the *strike* of the rocks and the course of the anticlinal wave, complicates the outcrops in a manner most perplexing to the geologist,

but in the highest degree advantageous to society, for it thereby brings to the mining surface over a much larger area the iron and coal upon which the prosperity of the community so much depends.

29. The so-called primary ranges of the Atlantic coast are in New England synclinal mountains of lower silurian, capped in the case of the White Mountains by Devonian, and so highly altered by injected volcanic rocks that they have not until quite recently been so recognized. One of the most beautiful applications of the *synclinal keel mountain structure* (explained above p. 136) has been made by Prof. Hall to this White Mountain range, in explaining by it their superior height above the Green Mountain range, inasmuch as the latter consist of silurian only, while the former contain within and on top of their silurian synclinals additional synclinals of the higher Devonian. The solitary elongated peak structure of these New Hampshire mountains, their abrupt terminations and lappings past each other, are all facts explained by their synclinal build. (Appendix VIII.)

30. On the contrary the South Mountain ranges, which further southward become the great Blue Ridge of Virginia and the Carolinas, are anticlinals (according to Mr. Trego and Prof. Bojé), and in a similar manner lap past each other, as they lift their backs composed of still more ancient gneissoid rocks through and above the plain of Potsdam sandstone, Trenton limestone, and Hudson River slates.

31. On larger fields of topography this law constructs the enormous echelons of the Alps and Andes. As laid down upon the common maps, these gigantic ranges form a single line, irregular in course but continuous across the continent. Probably none such exist on earth. The Alps are well known to be short parallel ranges, entirely distinct, lapping past each other, with high water sheds between them. Each range has its own separate axis and termini, but whether anticlinal or synclinal has not yet been clearly determined. The Andes

in South and North America are known to be constituted in a similar manner, but abound in volcanic outbursts.

32. c. **Mountains terminate when the rocks are cut off.**—In some countries this is a common event; in the Southern Appalachians it is the rule; in the Northern Appalachians it is the exception. A *fault* is a crack in the earth's crust on one side of which the rocks are thrown up and on the other side down, to the extent of an inch, or a fathom, or of even several miles. In the English coal fields these downthrows or upthrows are innumerable; Fig. 7, p. 46, represents them in the Richmond coal of Virginia. The topography of the United States has been too little studied yet, to say how far they are to become a common annoyance in the bituminous coal measures. Some which have been reported prematurely in the anthracite region are now on good authority and the best evidence, denied to exist; as, for instance, at the gaps of the Shamokin waters in Dauphin County, Pennsylvania, where gangways in the same bed on opposite sides of the gap have been found by careful instrumentation to lie in line; and the apparent thrusts of the Sharp Mountain must therefore at least in such cases be explained by oblique denudation. The celebrated Sharp Mountain Dislocation also is found by slope works to be no dislocation at all, but a plain synclinal, at least in the neighborhood of Pottsville. To say nothing, however, of the cross fractures of the Mohawk and the Hudson downthrow, one enormous fault does exist north of Mason and Dixon's line, and a little west of Chambersburg. The western side of the anticlinal Cove canoe has been cut off and carried down at least twenty thousand feet into the abyss, along a fracture twenty miles in length; the eastern side must have stood high enough in the air to make a Hindoo Koosh; and all the materials must have been swept into the Atlantic by the denuding flood. The evidence of this is of the simplest order and patent to every eye. The highest portions of the upper silurian, For. VIII, wall against the lowest portions of the lower silurian II. The thickness of the rocks between is of

course the exact measure of the downthrow; which is therefore twenty times as great as the celebrated Penine Fault in England. Yet a man can stand astride across the crevice, with one foot on Trenton limestone and the other on Hamilton slates, and put his hand upon some great fragments of Shawangunk grit, IV, caught as it was falling down the chasm, held fast in its jaws as it closed, and revealed by the merest accident of lying suspended in the crack just where the plane of denudation happened to cut it.

33. The phenomenon described above, is repeated in Wythe Co., Virginia, where, however, the out or eastern side of the canoe (in the Cove, it is the in or western side,) is opened by a similar but lesser fault, in which the Trenton limestone is thrown up against the successive edges of III, IV, V, VI, VII, and VIII. This, however, is the region of faults, most of them of immense size. A cross section of the crust as it would have appeared previous to denudation, and subsequent to fracture, is given in the following wood-cut, carefully drawn

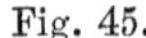

Fig. 45.

from observation, on the basis of the Va. and Tenn. R. R. map. The dark line shows the present surface approximately level, along which at intervals the lowest and the highest formations (II and XI) are seen braced against each other at the dislocations. This explains what has been said in the last chapter about the South Virginia coal. But, it is here introduced for another purpose, namely, to show how faults terminate mountains by cutting off their rocks; in fact, in doing one, they must accomplish the other also. A mountain of IV or a mountain of X, when it comes to where the fault enters it, must lose its central rib of rock, which has been replaced upon the opposite side of the crack by some very different, higher or lower, softer formation, one quite unable to form a mountain; of course the mountain ends. Up to this point it has been full formed, because well fed; it now perishes, so to speak, of starvation. The only possible chance it has of continuing in being—and one it certainly has—is the rarest conceivable, but one so extremely curious that we must hope some single instance of it may yet be found if only for an illustration of the principle. Should the downthrow be precisely equal in extent to the distance between two equally hard and massive formations, between IV and X for instance, then and only then would the extraordinary spectacle be presented of a mountain continuous beyond a fault,—more wonderful yet—of a mountain of one rock prolonged as a mountain of a wholly different rock. Should the record of such an event be anywhere preserved, its discovery would be likely to make geologists pilgrims, from however far.

34. It is doubtful whether the Blue Mountain which ends so abruptly S. of Newburg, N. Y., is thinned down or cut off by some downthrow connected with the great Hudson River fault. The chief evidence that the former is the case, is the alleged non-conformable superposition at Rondout, of VIII upon VI, and the fact that when a mountain is cut off on one side of a fault, it must appear again on the yon side, either further on or in arrear. A fine instance of this feature in the topography occurs in Pennsylvania, upon the Juniata River, below Hollidaysburg, see Fig. 46, where a small downthrow of about three thousand feet casts the mountain of IV, up *backwards, along the fault*, out into the valley of II. The scene illustrates, also, another fact of some importance, namely, that faults do not always necessarily make

Fig. 46.

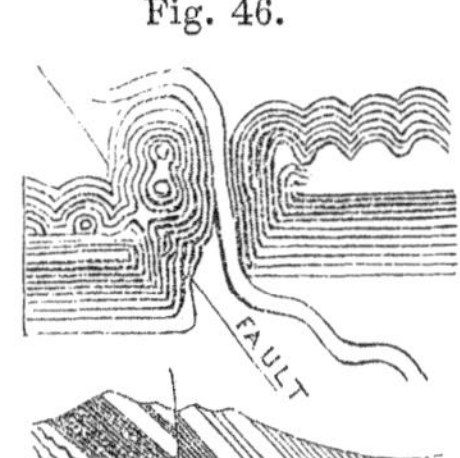

gaps. On the contrary, here where we should suppose everything prepared for a gap, there is a mere notch ; while the deep gorge, through which the whole back drainage came, and still comes, stands close alongside. I know of no explanation for this, but on the supposition that the direction of the diluvial forces was from N. W. to S. E. (the direction of the arrows), and therefore across the fault. The whole bearing of the flood would be round the end of the mountain at (*a*), against the point at which the river now flows through. The fault may be supposed protected by the crest of (*a*) when cut down to its present level, and only to have suffered from a side rush into the cove at (*b*). In fact, faults which are mere cracks, seem scarcely to have weakened the rocks ; the impetuosity of the attack seems to have overlooked all the minor differences of defence, so far as they lay to one side of its normal directions. It is, therefore, a question in geology for the mathematician why our gaps exist just where they do,—a question which opens a new world for the application of the higher mathematics to terrestrial forms.

35. **Eddy hills** and **Gaps** go together. The straight edge is found in nature so rarely, that it may be used as a characteristic. It characterizes *solid force.* The curve characterizes *fluid force.* All substances, it is true, are actually plastic ; even glass, copper, and sandstones of pure white quartz are flexible, and, therefore, have a certain fluidity among their atoms ; and the firmness with which the infinitesimal attractions of these atoms hold together in due form the concrete plastic chaos in all things glorifies the imagination with a heavenly view of the Almighty. But the attractions differ, and divide matter actually into *solid* and *fluid ;* so that we see the secondary forces constructing and destroying the *former* in right lines and the *latter* in curves. Where matter is very compact in its primary parts its secondary forms become rectangular. Hence, fissures with straight edge walls ; hence, the straight facet edges of crystals. On the other hand, where we have fluid freedom, the globe with its curved lines is produced.

Hence, the ball ores; the conchoidal fracture of fine mud rocks, once a pure paste, instead of the angular fracture of coarser and less fluid sands and gravels. Hence, also all the phenomena of subaqueous topography. Water is a thing of one idea; it describes curves, and can do nothing else. The curved points and cones of Appalachian valleys are, therefore, so many monuments of sea currents moving over its surface while submerged, or of a deluge poured across it after it had risen into the air.

36. The evidence of circular or **curvilinear cutting** surrounds the observer wherever he goes. It has too often been ascribed to the action of the present waters when the grade of the valleys and height of the water sheds and other important items have been overlooked.

But it is impossible to avoid the conclusion that these, the present waters, are the powerless modern representatives of those ancient floods which did the work. It is a theme of poets that such and such a mighty river has rushed with resistless energy against the mighty barrier of such and such a mountain—the Potomac and Shenandoah against the South Mountain for example—and breached a passage to the sea. It is, of course, contrary to the laws of gravity for a river to rush over a mountain top to *begin* such a work; and any preparatory accumulation of its waters in a lake behind would have left its marks, and could certainly have formed but one gap, if it could have formed any—which is doubtful. There are three breaches in the Kittatinny Mountain within a range of thirty miles. The middle one is back of Nazareth, and is called the Wind Gap, the mountain being cleft half to its base. Through the other two issue the Delaware and the Lehigh rivers. It is certainly an extraordinary fact—if the rivers have had anything to do with the construction of the gaps—that both Cherry Creek, which flows into the Delaware behind its gap, and the Aquanchicola, which flows down into the Lehigh behind its gap, *both head at* the very foot of the Wind Gap and flow *from* it both ways; while Brodhead's Creek, a third and larger stream, heads also at the same point, and makes a wider circuit to the north to reach the Delaware. The Delaware Wind Gap, which is of unique and remarkable beauty to the artist—to the geologist explains away more of these popular prejudices than any other local scene can do. Its location is at a shoulder where the rush of a super-continental deluge would be at its maximum. It furnishes an unanswerable proof that no water now running, made a gap; in other words, it proves the cataclysmic origin of every mountain gorge in the land; and it illustrates the circular cutting which peculiarly characterizes our topography.

One of the best exhibitions of this is seen in the following (Fig. 47), where the Susquehanna valley has been cut through

Fig. 47.

the nearly horizontal sand-rocks below the coal on its way out of the Alleghany plateau above Lock Haven. The cliffs of the semicircular escarpment in the distant background of the picture are nearly vertical for many hundred feet. The topographer finds himself perpetually recalled to the examination of the original set of the diluvial current, by notches and ravines coming down obliquely over one side of a valley and pointing towards some such concave wall of cliffs as this, hollowed out of the other side of the valley further on. The sweep of the flood against and around the shoulders of the hills is an endless study. Where two valleys coalesce in one, the evidence of fluid force becomes irresistible, not only in these alternate correspondences of the side walls, shoulder and hollow, but in curvilinear promontories, terminal peaks, and isolated eddy-hills. The following, Fig. 48, is from the valley of Pine Creek in Lycoming Co., Pennsylvania, where *two*

Fig. 48.

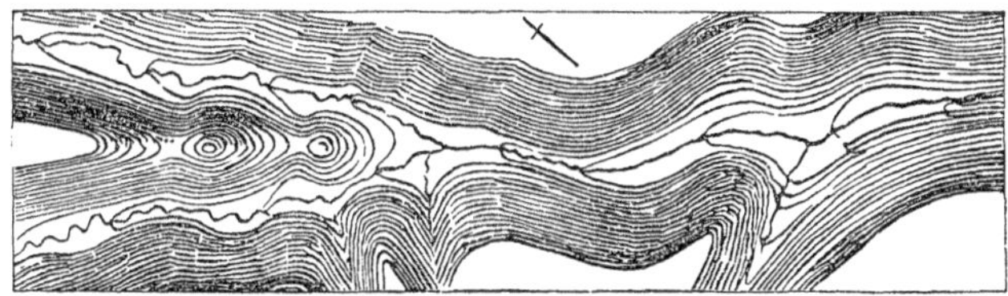

separate terminal eddy-peaks occur—which is a rare case. The currents which swept down the two branch valleys must have acted with symmetrical force, and the lower peak must have been created during the subsidence of the drainage waters, at the tail of the rush, after the eddy had shifted for-

Fig. 49.

ward from its first position over the higher peak. The *rationale* of the operation, however, would not be complete without the statement that the height of every such peak and the prime cause in fact of its existence in every case is to be found—as illustrated in the next Fig. 49—in some more refractory stratum of rock existing at that particular elevation. In the case before us, the upland is formed of the Great Coal Conglomerate cliffing out along the tops of the valleys' slopes. The valleys themselves have been excavated a thousand feet deep in the softer sands below. Among these softer sands there exist one or more very hard and massive strata, small plates of which cap these eddy-peaks, the only relics of the masses swept away. The fact is that these terminal peaks are essentially of the nature of the terrace, and might equally well be treated of under that head. It is evident that they ought to abound, as in fact they do, wherever the stratification is nearly horizontal. Whether their shape be straight or curved, will depend upon circumstances, chiefly upon the angle at which the one valley sets into the other. But that they obey a uniform law of conformation is visible in this, that under similar circumstances the same exhibitions recur. One of the most striking instances of it is the two eddy-prongs at the mouths of the two valleys of Broad Top, where Six Mile Run and Castillo or Sandy Run come into the Juniata. Here the dip happens to be 45°.

Fig. 50.

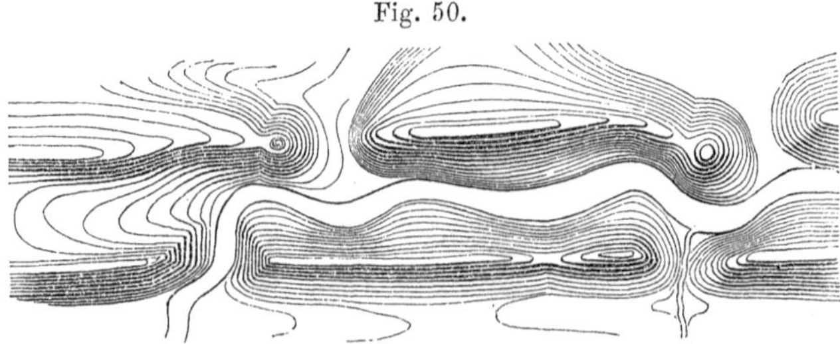

37. The **Isolated Eddy Hill** occurs wherever the angle of convergence of two equal valleys approaches 90°, especially among the horizontal rocks. Instances are very numerous. In

the valley of the North Branch Susquehanna there are several prominent ones, as at Oswego and Binghampton, and a less perfect one in the centre of the great Tioga River delta above Athens. The large map of Alleghany Co., Pa., shows two very fine examples; one in the centre of the Pittsburg plain, where the two great valleys of the Monongahela and the Alleghany come together to make that of the Ohio; and the other where the valley of the Chartier Creek comes into the last. On the former has been built the Alleghany city seminary. It is not a perfect cone, but somewhat longer than it is broad, and has almost a double summit. Its parent eddy revolved in an ellipse. The other is high and long, rising from the centre of an island in the Ohio. On the Beaver River, in Western Pennsylvania, is a very fine one above New Brighton, of which the following Fig. 51 will give some notion.

Fig. 51.

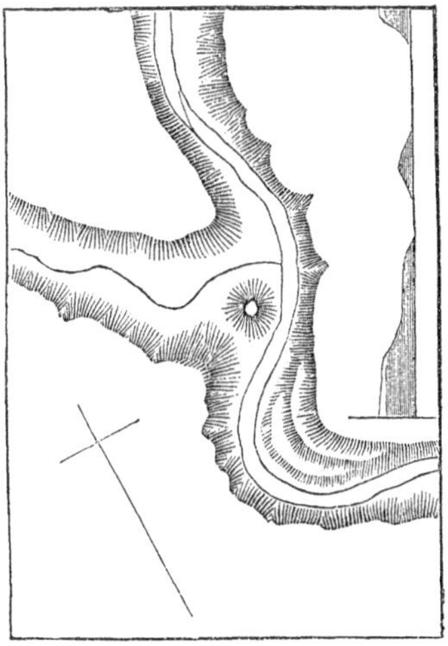

The cutting here wide round the eddy hill and against the cliffs of black and olive slates capped with quarry stone, is striking. The number of these hills in the west is very great and may have fortified that traditional prejudice for artificial, sacred, and sepulchral mounds common to our aborigines with all the nomadic races of the Old World. Some of the largest of the so-called Indian mounds are in fact eddy hills.

38. Three forms of this phenomenon must be alluded to which possess peculiar interest; *i. e.* the eddy hill

occupies **three** distinct **positions.** Wherever the diluvial current set broadly against a gap, as in the annexed cut of one of the Bald Eagle gaps, near Bellefonte, a small circular eddy was, of course, produced, which left its hill *in front.* Again, wherever a gap was double, as happens to be the case with the same mountain near Williamsport, at the entrance to the Nippenose valley, Fig. 53, where the rocks stand at 50°, an eddy hill has been formed in the heart of the gap itself, of a perfectly symmetrical shape, but (as in all other cases of high dip cutting) attached to one of the four mountain ends. Again, *behind* a gap, when circumstances proved favorable, an eddy was formed, which left an eddy hill. In the middle of the Hudson River, at the entrance, and at the exit of the Highlands, stand two fine island examples of the first and third of these three rules. The eddy hill is sometimes partially, and probably often totally submerged, forming peak islands, or circular shoals. And sometimes they are partially submerged, not in water, but in the delta mud of rivers, or rise from the plane surface of broad diluvial washes in valleys, the depths of which are sometimes very great.

Fig. 52.

Fig. 53.

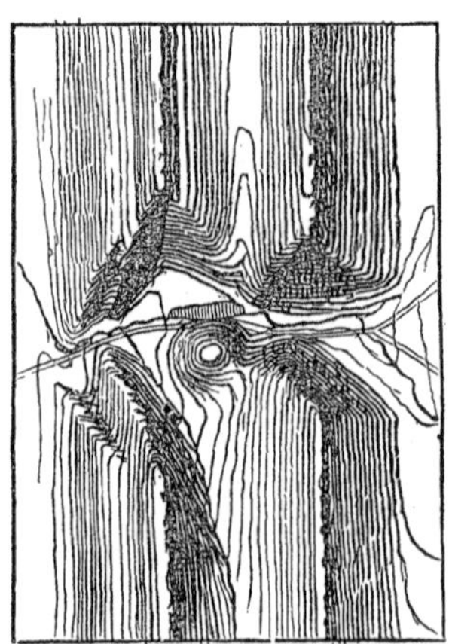

39. These details might be indefinitely multiplied, but the object of this manual is to illustrate broad principles by a few original authentic observations, and rather to suggest the

direction in which investigation should proceed than to present a summary of the results which the author's own investigations have produced. He may be permitted, however, on the score of the experience of years of labor in the field, to warn the numerous neophytes which yearly claim initiation into the Wernerian and Huttonian mysteries of this our genial science, against hasty generalizations of every kind. Millions of facts are barely sufficient to sustain some of its sublime hypotheses. However much one man may suffer, do, or tell, his name and contribution must be finally absorbed, and disappear within the rising tumulus of facts and theories for the completion of which he ought alone to live. He builds his own nameless

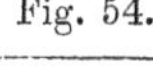

Fig. 54.

tomb. Able to do so little to prove what he thinks true, he should believe but little to be true, and believe absolutely nothing that he cannot prove; he ought to hold his dearest prejudices, his most brilliant suggestions, with a grasp as delicate and anxious as the bubble from his pipe before it is to burst, or his infant at the hour of its birth. The same apparent causes he will find producing the most various effects, and anon, apparently identical phenomena reappearing under utterly different circumstances. What traveller acquainted with the look of the exit gorges of the Alleghany Mountain in the New World, cut down through nearly horizontal late Silurian and

Devonian rocks, would imagine at the moment when he obtains his first view, from the summit of the Teufel's Mauer, of the

Fig. 55.

celebrated Gorge of the Bode (Fig. 54), as it issues from the Hartz Mountains in Northern Germany, that it expressed the rim topography of a plateau of aboriginal altered slates, set at nearly vertical angles? The outside resemblance is perfect; the inside structures could hardly differ more. On the other hand, what tyro in geology would not stare to be told that the stratification in Figs. 55, 56, of some of the pulpit rocks of the Juniata was *not* vertical, but horizontal? Surely, nothing could be more deceptive. Every experienced observer

Fig. 56.

will confess that he has not unfrequently met with exposures which he absolutely could not read, to which he has returned again and again without success, and one or more of which perhaps remain to the present hour a sort of spectres to mortify the pride of his science, and torment its waking dreams. A dip like this, for instance, in Fig. 57—what is to be done with it? It is a sphinx's oracle; and practical science is more

Fig. 57.

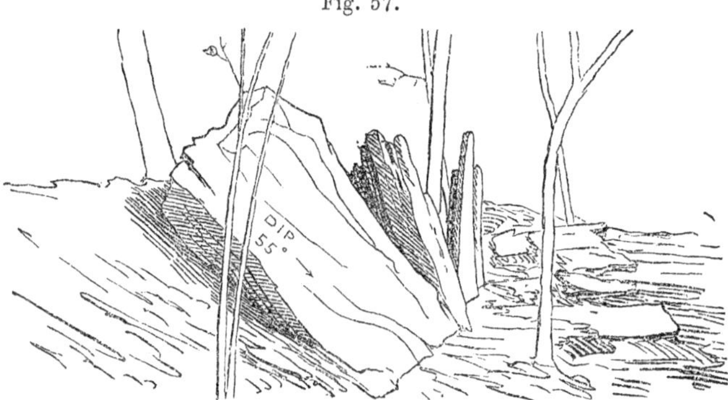

garrulous in such than an old woman, and sometimes more tantalizing than a shrew, concealing precisely when and where we wish her to reveal, and revealing under forms so distorted and absurd that faith, enthusiasm, and patience alike give way.

40. The sketches of the pulpit rocks presented on the preceding page bring up a final element of topography not yet discussed. Hitherto we have considered the mountain as constructed outwardly upon the mould of its internal structure—its stratification, or planes of original horizontal deposit. But subsequent to this its first formation, in piled-up horizontal layers, the earth crust has been *cross crystallized*, or subdivided into *laminæ across the layers.*

Lamination and *Stratification* are phenomena equally essential to the present structure of the crust, but essentially opposed to one another, and never more opposed to one another than in Topography, the discussion of which would therefore be very incomplete unless both elements were valued equally.

41. **Lamination** is perhaps best seen in old slate formations, such as the Cambrian or Silurian, the Hudson River slates, where it obliterates the original planes of deposit and splits up the mass in a new direction into roofing slates.

Fig. 58.

The cause of lamination is supposed to be a heating over of the first laid parts of the crust, and its crystallization under galvanic and magnetic supervision. Whether any extraordinary degree of heat was used or needed in the process—whether mere solar heat, or the slow radiant earth heat, or the propagated heat of volcanic zones did not suffice to produce it in the older rocks is still a question. But we know that its principal home is in the older formations, and in the neighborhood of traps and other volcanic rocks. Here it almost obliterates the marks of original sedimentary deposit; for the very genius of the crystal is recreative, beginning anew the arrangement of particles and ignoring all former statical relationships. On the other hand it commonly fines away and almost disappears in the undisturbed and late formations, where in the same proportion the stratification intensifies its markings. We call rocks *altered* which have lost the latter and gained the former structure; which have lost their *dip* or bedding and gained a **Cleavage.**

42. But lamination or cleavage is not confined to altered rocks; nor does it in fact from any rock, new or old, *entirely* vanish. One of its most curious phenomena is its excess in certain strata. Passing for instance through new railroad cuttings in coal measure shales, the eye will be attracted to thin sheets

of clay, ferruginous perhaps, or calcareous, or magnesian, or in some other way differing from the mass above them and below, but most of all in this, that they are closely cross cleft, or prismatized. Perhaps they are not three inches thick, yet they may run a hundred yards, projecting like the edge of a book cover from the leaves along the slope, and marked by this peculiar lamination. The same is true of coal; some layers in every bed are prismatic or laminated and the rest merely stratified. The vertical lines and planes in the pulpit rocks above are lines and planes of *cleavage*, while the lines and planes of stratification may be observed crossing them horizontally. In the views of these rocks given on pages 61, 62, the two series are equally apparent, and it is visible at a glance that these were vulnerable along their stratification and along their cleavage too. This is the point in Topography which is here insisted on. It holds good universally; for all rocks are both bedded and cleft, both laminated and stratified; and the action of the diluvial force could not but be governed by both facts. The stratification ruled in the general, and the lamination in detail. By the former the mountain form as a whole was obtained; by the latter its incidental features were modified. In alpine summits the ridge shows the strata, the aguilles the laminæ. In Appalachian topography the dip gives the slope and position of the cliff as a solid outcrop, but the cleavage determines the aspect of the individual crags which compose it. The pulpit rocks *without cleavage* would have been an uninterrupted wall, slowly rising for miles along the Warrior Ridge; *without stratification*, the pulpit rocks would have formed a crenulated terrace like the gopher work upon a collar; with both combined they form a row of fantastic and imposing columns, of every variety of size and shape. The rocks of the Saxon Switzerland upon the Elbe above Dresden, so celebrated, owe their existence to a similar combination, but under very different circumstances. (Appendix IX.)

43. The **Law of Lamination** is not well understood.

We know that heated plates crystallize in cooling *crosswise*, or perpendicular to their chilled surfaces. In Lava currents this develops the basaltic column, as at Staffa, in the walls of Fingal's Cave, the floor of the Giant's Causeway, the cliffs of French Auvergnese lavas, the East and West (trap) Rocks of New Haven in Connecticut, and all over the world. In iron it is seen upon the broken rim of a railroad wheel. We may therefore suspect that the extra cross cleavage or prismatic structure in certain layers of coal and clay and stone may be due to a difference in their power of conducting heat, and therefore in their rate of cooling. (Appendix X.)

44. However this may be, the fact upon the surface is that fine rocks like clay slate part in frequent and close cleavage planes, breaking up the whole mass into pencils; while coarse rocks, sandstones, and conglomerates part at wider intervals into more extensive cleavage planes. Whatever our views of the rationale of the process we see that there is some law according to which the size of the crystals formed by cleavage is in proportion to the massiveness of the strata. There is, therefore, some fixed relationship between the stratification and the lamination; and we know before that the massiveness of strata holds a similar fixed relation to the coarseness of their materials. Sand-rocks, therefore, as their layers thicken, break into huger cubes; as they thin down to flags they commonly—not always—cleave up into smaller blocks. The exceptions to this rule are numerous and will afford fine exercise for the synthetic genius of some future day. It has not been sufficiently recognized that among the many characteristic distinctions of strata, is also this, that some one stratum in a formation is peculiarly disposed to cleavage or the prismatic structure; if a clay, into pencils; if a sand, into clubs or bars of rocks; and that not locally but universally; not therefore as an altered, but as unaltered rock. There is a portion of the Clinton Group which everywhere conceals its stratification under a cross lamination to such an extent that it is often impossible to tell how it dips. The finest show of this it makes, perhaps, is in the gap of

Broadhead's Creek behind the Delaware Water Gap. There is a sandstone in the coal measures which breaks habitually into immense pencils. The layers of the Conglomerate throughout the Broad Top coal region confound the most experienced observer with their cleavage planes and present some of the most notable examples of the effect of these cleavage planes in tempering the local topography.

45. **Lamination in Coal** has been alluded to. It is common to coals of every quality, both anthracite and bituminous; it splits the harder varieties up into cuboidal or rhomboidal masses and the softer into minute prisms; it guides the miner in his processes of picking and blasting, giving him what he calls his heading or butt to the coal, along or across which his blast or his pick does more or less execution. He gives, in fact, a rude but true expression to a further law in lamination, namely, that it has not one plane but three, one of which, however, always preponderates. Its crystals are completed by itself; stratification may mar but does not help to make the crystals; stratification is always a fourth and as to the crystals an abnormal element of division. It is easy to see that if the force of lamination northeast and southwest preponderates over the force of lamination north-northwest and south-southeast, we will have a greater number of subdivisions or clefts in the former direction than in the latter; the whole mass will point its principal lines in that direction, and its flags will stand up with their ends that way, and the denudation will have left a cliff ragged in that direction with a smooth wall towards the other quarter. This is what the miners mean by the ending or heading of their coal; and this is what produces such remarkable effects in the topography of the conglomerate and other massive sand-rocks of the coal.

46. There is a **peculiar crystallization in coal** which the author has never seen but once, and has carefully attempted to reproduce in the following sketch. It is to be seen at the mouth of the great opening on the Cumberland Mountain in Tennessee, about ten miles north of the summit of the Nash-

ville and Chatanooga Railroad. It must be very rare, and resembles nothing but the curious and unexplained concentric "crucible crystallization" called *tutenmergal* by the Germans, seen in certain clays, especially those below the Freeport

Fig. 59.

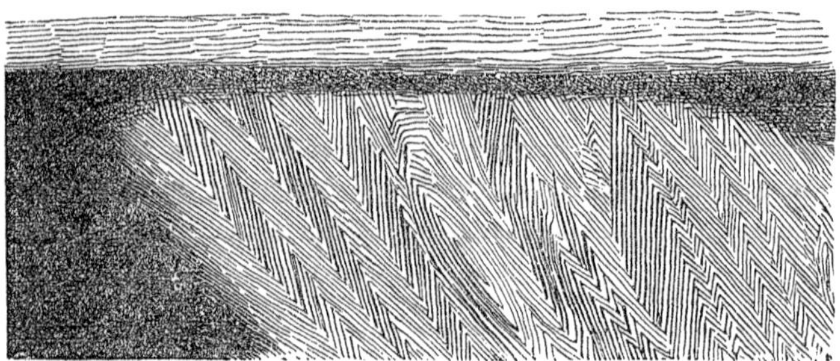

coals. (See *Owen*, page 123, 127, &c.) As it was seen only at the mouth of the opening, where it had little cover and was not a hard good coal, it might be due to the infiltration of clay, if that be possible. But the author decided in his own mind at the time that it was probably an interference of two or more sets of lamination planes of equal force meeting at low and variable angles. It is a curious coincidence that within this opening he also found an almost equal rarity, namely, a true false bedding or oblique deposition in the coal itself (Fig. 60), with every essential point made clearly out, the feather edges below, the truncations above, and the curved lenses of slate. It is clearly a swash of coal not differing otherwise from common beds; in confirmation of which it is only necessary to add that the same bed is here twenty feet thick which is not known to exceed six feet any-

Fig. 60.

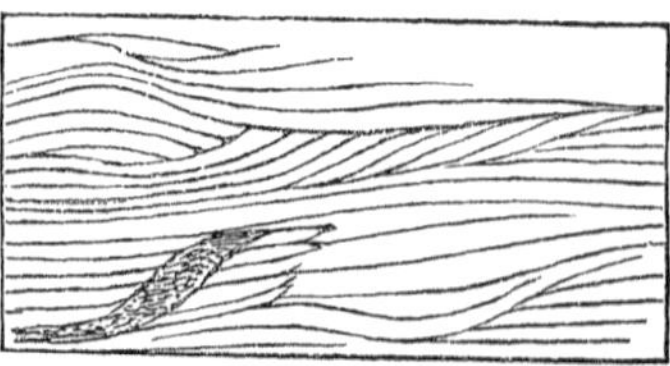

where else upon the mountain. We are perhaps too apt to assign every irregularity in the thickness of a coal bed to the sliding, crushing, squeezing, and bulging action of the upheave. Our study of the over disturbed anthracite region has prejudiced us in favor of this view. It is undoubtedly correct. Such immense masses of coal as immediately succeed nipped and faulty spaces in the gangway, such as Whelpley's earlier diagrams will show in the Final Report, can be due to no agency but that of a sliding crush propagated through the rocks, coal and all, in a direction diagonal to both their original deposit and their subsequent upheaval, from which crush the coal would suffer most and be squeezed, sometimes in company with its soft floor and roof, from the thin places into the thick. But other most important and far earlier irregularities in the coal beds themselves undoubtedly took place, most of which were of the nature of false bedding or drifting of the coal in a semi-fluid state, as shown by the last diagram, and everywhere proven by the shape and position of the parting slates which so frequently come in and go out from the beds. These were originally layers of sand or mud swept in upon the vegetable soil, and often partially swept off again before the next layer of the coal was deposited.

47. The fluidity of coal under certain conditions of deposit, has been most curiously demonstrated by a late discovery upon the Mississippi, made by Prof. Hall, who has courteously permitted the author to allude to it in advance of its proper publication. Prof. Hall found a thin layer of coal exposed in a quarry *beneath* silurian limestone, and upon clay; above the limestone lay the lowest of the coal beds of the carboniferous. He sought in vain for a solution of the mystery. At a subsequent visit he found the quarry greatly enlarged and the presence of the coal explained. The clay was seen to occupy an ancient cavern in the limestone; a vertical crevice explained its entrance to the cavern; the solid coal above had furnished a liquid coal which descended the crevice and floated upon the surface of the clay at the top of the cavern; a *leader* of coal

in the centre line of the crevice joined the true and false beds together. How much of the coal of the earth has been deposited in this floating condition we cannot tell, but certainly enough to make us cautious what we take for granted in the *modus operandi* of its sediment. Much that is called lamination may turn out to be oblique bedding, and much that has been called "crush" and "trouble," may be original "pocket" laying.

---

48. **The Theory of Denudation** is a subject of such general discussion, occupying so much of the best thinking of the best geologists now living—has been looked at from such opposite stand points, and decided by reference to such controverted facts, that the author feels permitted to treat it as a discussion essentially premature and to state only the ground over which it moves. That it lies back of all our practical rules of geological topography is evident at a glance, and therefore something had to be said of it at the beginning and much concerning it assumed in the course of this chapter. For, like every other theory, it has both its facts and its hypotheses; and while its facts are necessary to elucidate the phenomena of the earth's surface, its hypotheses scarcely affect our practical use of them at all. Its facts, however, are not all of them universally accepted; the very first and grandest of them meets with a cold reception from the majority of thinkers. The rush of an ocean over a continent is thought to be too improbable a fact to be more than a poetical vision and therefore is thrown aside among the hypotheses not yet dignified into theories; whereas to the field-worker it leads off the whole procession of his facts, and is indispensable to the exercise of his sagacity at every turn. It is the first point, therefore, to be settled in the study of topography, and has been assumed as settled in all that has been said above about the mountain, its cross section, its crest, and its end, the gap, the bowl, the canoe-cove, and the eddy-hill.

49. The arguments of force in proof that a cataclysmic deluge wrought out our topography, have been introduced into the body of illustrations on the foregoing pages. Conversely, those that are thought to prove that quiet, cyclical, ocean tidal or current action could not or actually did not work it out, have also been more or less clearly stated. The case of the Wind Gap has been cited as conclusive evidence that the present waters have had no part in the matter of denudation so far as our gaps and gorges are concerned. But there remains at this one point something to be said which will reinvolve all the others.

50. **Cataract Ravines**—how were they produced? We know that the great lakes of Erie, Ontario, and Michigan are not synclinal basins, sinkings of the crust, and inflowings of the sea, but outcrop valleys of *denudation*, scoured out to a great depth in the broad, flat, slate layers of the immense Hamilton group, VIII, and of the Hudson River group, III. We know that they would have been drained like all the rest of our valleys of VIII and III, had a gap or gaps deep enough been made to put their drainings in communication with the lowlands on the sea. We know that the power which excavated these lake valleys must have been great enough to have previously carried off all the formations which outcropped above them, and was evidently great enough to make a vast system of valleys, long, strait, narrow, and deep, pointing north and south, and northwest southeast, across the great upland of the Alleghany coal; valleys which are all thorough cuts from the lake country to the Alantic seaboard; valleys with high cliff margins and low water-sheds; valleys large and deep enough for great rivers, but the most of them occupied by insignificant streams; valleys which never could have been excavated by their present waters, for wherever the rocks are soft enough their beds are engraven so deep that lakes have remained in permanence upon them. Such are the valleys of the Upper Delaware, the North Branch, the Lycoming and Loyalsock, Pine Creek, the Sinnemahoning, the Genesee, and

Alleghany, the valleys of Seneca, Cayuga, Canandaigua, and Otsego lakes, and many others of names less known to fame. These could not have been excavated by their present waters which flow through them in *alternate* directions. It has frequently been said that there are traces of the action of their streams upon the high mouths of some of them where they intersect and look down upon the great cross valley of the Mohawk; but these marks are vestiges, not of the Susquehanna and its branches once flowing northward, but of the deluge which cut the Helderberg escarpment down to its present height, notched its rim, commenced the southern valleys at these notches, widened and deepened them under the long strait lakes in the soft shales and sands of VIII, cut down the Devonian carboniferous upland of Pennsylvania and Virginia to its present moderate height of 3000 feet above the sea, and continued across it in long lines, the sharp, strait, saw-cut valleys on to the Atlantic coast. That this was not done under water it has already been argued is proved by the absence of all later oceanic deposits. That it was done in the air is proven by the angularity of the fragments it has scattered. That it was not done by ice is proven by the absence of moraines and by the shape, direction, and position of the scratches which are left. That it was done by a moving ocean is presumable from the force required to break and remove vaults of earth crust from two to seven and even twenty miles in height. That it was done from north to south is shown by the fragments of the Green Mountains on the flanks of the White Hills, by the rocks of New York in the valleys of Pennsylvania, and by the topographic cuttings. And that it was done at two distinct eras, one before and the other after the New Red Era, is proven by the two facts, first, that the New Red is deposited in *estuary valleys of denudation* already formed in the Palæozoic Silurian formations, and second, that the surface of the New Red is denuded in perfect harmony with the surface cutting of the older rocks beside it, as if it formed a part of them. This *double era* is the only fact which looks towards suboceanic denudation. All our skill and patience will be tasked to determine, since there have certainly been two eras, whether there have not been more—perhaps many more—and when. In fact we already know that this denuding process went on at many times. The formations in one part of the country do not succeed each other as in the other part. Gaps occur in the series of the rocks. In Pennsylvania the Niagara group is wanting. In the west **coal lies upon silurian**. It is not enough to say that intervals of *dry land* occurred. We have perfect non-conformability; coal lies on the *upturned* edges of rocks far lower in the series. There were repeated denudations, therefore. The topography of the crust did not become so at once. To what extent, then, at the first time, and at the second, and so on to the close? When was the great process perfected? How much must we diminish the force of the agency to suit the diminution of the act by subdivision? And how far can we discover in the conglomerates of different ages the vestiges of successively preceding cataclysms and fresh modifications of the external surface; finally, how far were the first of these modifications subaqueous and the final ones subaerial? A volume may be written on the facts already collected to answer the last of these questions; but very little has yet been found to base replies to the first. They are all involved, however, in the question before us—how and when our *Cataract ravines* have been produced.

51. **Niagara Falls.**—Much poetry has been expended upon this grand scene, strangely enough, in behalf of chronology, the most thankless and stolid of the sciences, endowed with all the precise angularities of mathematics, and with the *mouton qui rêve* profundity of history. There no doubt *is* a chronology of the prehuman earth, but as yet, all attempts to obtain a scale for it have been abortive.

The leaved shales and slates furnish no datum; because it is impossible to decide whether they were deposited one at every tide, one at a storm, a freshet, a year, or all together and afterwards laminated horizontally by percolation.

The size of fossils is no guide to the age of the rocks which inclose them, for we know not the rapidity of their growth, nor their organic methods of self-distribution through a given vertical height.

The decomposition and revegetation of lava surfaces has been tried, but has failed to yield any but the most discrepant results, ranging between a hundred and a thousand years, differing for different lavas and for different climates.

The swamps of deltas and successive overlying submerged forests and peat bogs fail from the absence of individual dates from which to start and from our ignorance of the rate of growth of various vegetable forms.

The problem succeeds no better with the destruction than with the formation of the rocks, either in the lava regions of Central France, in the curious horizontal valleys of the Hindu Ghauts, or at Niagara; and that, for want of the most important element of the calculation, namely: how far the process of destruction was accomplished at one time and by one effort, and how much was left to the slow, and, therefore, supposed mensurable action of the ordinary forces which we know, itself however also most imperfectly estimated.

To elucidate this view of what is an important point in speculative geology, but a much more important and altogether vital one in structural, or topographical geology, the following reduction of the New York State measurement by Professor Hall, of the Falls of Niagara, is here given. (Fig. 61.)

Fig. 61.

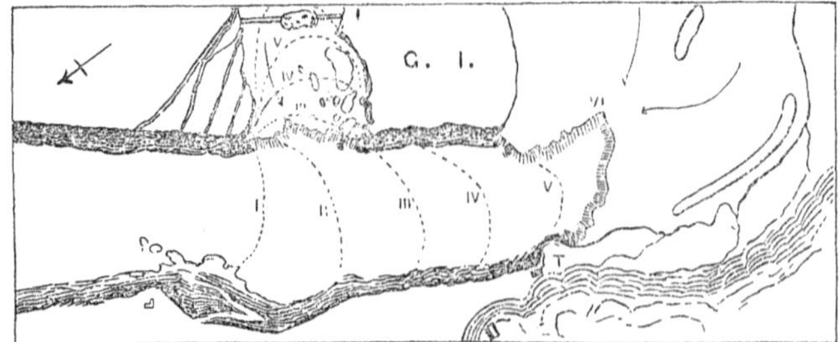

It will be seen, by this sketch, that the American arm of the river falls over the *side* of the chasm, while the British water runs in a larger body round the head of Goat Island, and plunges into the chasm at its end. The difference exists in the kind of rock over which both fall; therefore, there is no reason for the destruction of the one dam faster (proportionally to the different quantities of water) than the other. Prof. Bakewell, of England, thought he was justified, by reference to Father Hennepin's sketch, in 1678, and Kalm's, in 1751, and from reports of old men living in the neighborhood, in assigning 3 feet per annum as the rate of retrogression of the falls. Sir Charles Lyell, after two visits to the place, in 1840–41, apparently for no precisely determined reasons, but under a general impression that 35,000 years fitted better into his theory of moderate geological energies than 10,000 would, thought he ought to reduce this rate to one foot per annum; and although his opinion was avowedly the merest inductive suggestion, it has since been assumed by his world of readers as a positive fact, and accepted upon the eminent authority of his name. The first attempt to reduce the elements of the case, by a truly philosophical analysis was published in 1854, in Neuchatel, by the Swiss geologist Desor, who brought to its investigation what had hitherto been wanting, viz: an intimate acquaintance with phenomena characteristically allied to its own, obtained by a long and familiar personal knowledge of the waterfalls of the Alps and accurate measurements of the movements of glaciers. Feeling that such rapidity of action by water on rock as Bakewell's and Lyell's calculations went to show was unnatural, he argued that, even if we grant the indentation in the American fall (*a*), which is only 132 feet back from the strait line (*a*, *b*), to have been made since Father Hennepin's time, 174 years ago, it only gives a *partial* recession of 9 inches per year, and an average of only half that, or 4½ inches per year; and that to fix the straight edge of the American Fall at that date, necessarily supposes that at that date also the English fall was flush with and a prolongation of the American fall; otherwise, what was the American fall doing all the time that the English fall was cutting its way back from this to any other preferred position? Did the American fall previously *project* into the chasm? That would be impossible. Father Hennepin's picture, on the contrary, exhibits the two falls as far asunder in his day as now; we must, therefore, suppose the indentation in the American fall to be "infinitely older," and of course its rate of wear and tear infinitely slower than 4½ inches per year. Instead, therefore, of Bakewell's three feet per *year*, Desor prefers much to imagine three feet per *century*.

Had Prof. Desor, however, previous to his visits to Niagara, had the opportunity of studying the dynamic geology of the Jura with the key to it afforded by Appalachian Topography, as he has since done, or had this topography extended itself, with all its distinct and elaborate detail into the districts studied by the New York geologists, the question would have been long ago raised by them whether in fact after all this speculation upon the *rate* of retrogression, any *such* retrogression has ever taken place. M. Desor came nearer the truth, because his experience in the Alps led him to feel how refractory such a plate of limerock must be, over which all the waters of the lakes might storm for generations without producing a change determinable by ordinary triangulation. The only change in the falls since Father Hennepin's day is the diversion of the small oblique cascade from table rock; and the occasional droppings of the ledge, amounting perhaps in all to a few hundred tons a year, is algebraically nothing to the bulk of a single yard back from the face, a thousand yards long, of a stratum thirty yards thick, containing, therefore, fifty thousand tons of stone. Taking the recorded falls of table rock as a basis of calculation, the British fall could not have touched the American fall at right angles less than **six hundred and fifty centuries ago.**

52. But the question is now in place which shows the fallacy of all this reasoning from the doubtfulness of its starting-point.

What was the American Fall doing all these sixty-five thousand years? Why does it in fact at the present day still fall over the front edge of the chasm, while the British Fall is supposed to have cut back from it six-hundred yards and more? In fine, why do not *both* falls, instead of only one, fall into the *heads of* two *cul-de-sacs* of their own manufacture, and of different lengths in proportion to their different volumes of water? There seems to be but one reply to this: the falls have *not* excavated the chasm. This was then done long before the waters of the west obtained this channel. When they made their outlet here they must have found the chasm of the Niagara River already excavated and adopted it as their only available opening to the St. Lawrence; since when they have only modified its edge by breaking it into two horseshoe curves proportionate to their divided volumes. The question of chronology then passes over entirely to the discussion of another set of facts, viz: those connected with the change of western drainage from southward into the Gulf to northward into the Atlantic, and of those far more ancient facts connected with the excavation of the ravines of the Genesee, the north and south valleys of the New York Lakes, and the thorough-cut valleys of the Appalachians.

53. Even if we were to grant an actual retrogression of the Niagara Falls, it would be impossible to avoid the discussion of this preparatory denudation on a grand scale, which Sir C. Lyell acknowledges by allusion but does not open up. And where did the Falls begin, would continue to be as much a question as ever, all calculations of time based upon its rate wanting one of their two indispensable factors.

54. This is set in the clearest light by certain facts; first, the curious right angle zigzag course of the Niagara River, bringing it into close relationship with all our Appalachian rivers, and so far fatal to the notion that it constructed its own channel (see Fig. 62);—secondly the fact of its following a valley within a valley, a lower in an upper, which agrees again with all our Appalachian topography;—but especially the existence of

the St. David's outlet about which Sir Charles Lyell has much to say in his second volume and much that is extraordinary, while he omits the most important item in the account, and that which suggests objections to the rest. The Valley of St. David's is in fact a *straightforward* and original mouth of the Niagara Valley, from which the actual valley takes off at right angles at the whirlpool. This valley is described as filled up with diluvium to a height of 300 feet; that is, flush with the table-land; but it is not stated why in the opinion of those who assign the origin of the chasm to the abrading action of the Niagara waters these did not attack this loose and portable mass of gravel in preference to cutting a toilsome way through the ancient massive limestone of the cliffs and so issue upon the plain of the Ontario at St. David's instead of at Lewistown. (*See* Appendix XI.) Especially is this question pertinent since the gorge at St. David's is at the lower end, *not* filled up with diluvium, but open and deep, offering no dam to the waters of the Niagara at the time of the supposed commencement of the Falls at Brock's Monument above Lewistown; at which time the Niagara waters would have been of course so high as to be backed over all the country of Western New York and Southern Canada which lies behind the escarpment of Limestone, and would inevitably have rushed down the St. David's Valley and the present route of the Welland Canal in preference to cutting for itself a passage over the cliffs scouring away any diluvium which happened to be in its course. It is in the highest degree therefore probable that no such damming of the waters ever took place; that the excavation of the Niagara Chasm and of the St. David's ravine was simultaneous and at

Fig. 62.

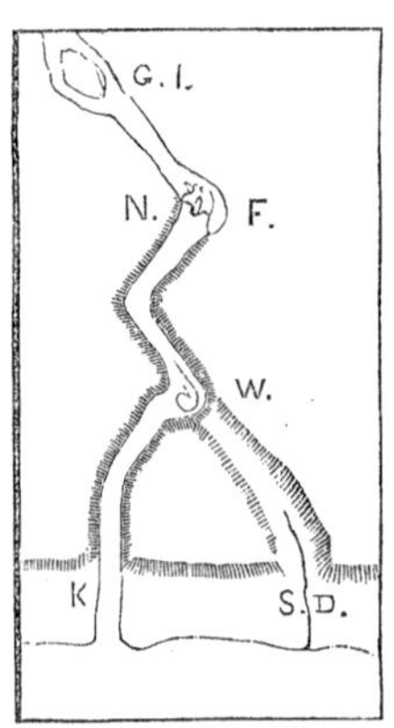

the time of the formation of all our north and south thorough-cut valleys and ravines with their outcrop terraces, water-sheds, precipices and cataracts; and that the Niagara Chasm, happening to get excavated a little deeper than the other, that fact was enough to decide that the subsequent drainage of the lakes should adopt its channel in preference to the other. Both valleys no doubt were originally filled with diluvium more or less; the one adopted by the river would alone be scoured out; and yet the remains of the Niagara Chasm terraces of diluvium are sketched in Mr. Hall's map and described in Mr. Lyell's travels. What he says of the different width and shape of the St. David's ravine seems at first important, but unless a much more thorough sounding of the diluvium by wells or auger holes were made than has been or indeed could be made except at great expense his conception of the terraced form of the cliffs of the St. David's Ravine beneath the diluvium may be and I have little doubt is entirely erroneous. The upper member of the limestone which forms the rapids of Niagara would anywhere fall into terraces, but the ninety feet mass below would everywhere form cliffs. The presence of the hill of sandstone struck in boring in the centre of the St. David's Valley is precisely analogous to the rise of Goat Island in the centre of the Niagara Chasm. More than this—the author ventures to predict that if ever borings be sunk in the centre of St. David's Valley near the whirlpool and back from it say half a mile such borings will strike rock far short of the supposed depth of the diluvium, the face of which will be then seen to be a mere mask or bank, a wash of heterogeneous detritus in a deep cirque backed by semicircular cliffs of limestone never yet removed, and that the true reason why the waters did not adopt that channel in preference to the one they now pursue is to be found in some comparatively high limestone water-shed concealed by the diluvium;—a reason however that would be of no force if we started with the supposition that the original dam of the lakes was the summit of the escarpment of limestone at Brock's monument.

55. It is greatly to be wished that Mr. Logan, chief of the Canada survey, and Mr. Hall, who alone continues officially to represent the New York survey, would institute a topographical examination of the crest line of this escarpment and the water-sheds immediately back of it. This would be decisive. Precisely upon such points as this hinge the rival theories of ancient cataclysmic or furious subaerial denudation and of quiet, usual, gradual, suboceanic shore fretting action such as Sir Charles Lyell describes in his discussion of Niagara and favors in all his books. To one who has studied the high mountain topography of the interior, far from the sight and sound of present oceans, in the bosom of which the largest rivers lose themselves as insignificant brooks, destitute of any vestige of littoral action, possessed of no tidal terraces, no osars; no "parallel roads," no lacustrine deposits, not a trace of those horizons of successive emergence which we would expect to see, but exhibiting everywhere with a magnificent unity of parts and harmony of detail one law of denudation, uninterrupted in its paroxysms and essentially instantaneous, the idea which anticataclysmists cherish of the gradual wearing away of our valleys and our mural escarpments under water, simply excites astonishment. For how is it possible that the tidal currents which wore away the edges of the Catskill, and the Alleghany, could fail to fill with its detritus the valleys over which they must, in doing this, have swept,—in which valleys not a vestige of any such deposit has ever been suspected? This alone seems to set the whole subject at rest. There seems absolutely nothing after this for discussion. Until this first position is assailed all discussion seems to be reversed. For the first, the broadest of all topographical phenomena in Appalachian regions, is the cleanness of the surface, the completeness of denudation and the entire absence of all slow contemporaneous deposits; and in the stead of such, sheets, banks, and eddy-mounds and stripes or threads of angular fragments, the products of the power which did the work, and the evidence that it was done with infinite force and speed, and ceased forever.

56. From all that is here said, however, must be excluded that acknowledged latest of all deposits, the **Drift,** with whatever surface cutting may be hereafter proved exclusively connected with it. Around the whole northern cope of the planet an infinite number of lines of transported materials radiate towards the south, the south-southwest, but especially the south-southeast. They form a sort of fibrous epidermis for the earth, coarser at the north, and growing finer and finer to its southern edge,—if indeed it may be said to have an edge; for although the streams of boulders stop upon a line irregularly waving between the 40th° and 50th° of N. latitude, single blocks advancing as far south as Leipsig and Cincinnati, yet the whole temperate zone is covered with a general sand and loam and with local washings towards the south which can hardly be assigned to any other time or cause than that of the drift agency. What this was, is the great geological question of the day. One party holds it to have been an oceanic iceberg movement, such as is still covering the Atlantic bed with northern blocks and sand. Another party votes for a universal sheet of snow and ice, pushing forward its diverging lines of moraine matter slowly over the highest mountains as easily as across the lowest plains. A third sees but a repetition of the ancient cataclysms, this latest of all the deluges, which breaking up the northern fields of ice, and sweeping across the surface all its loose or detachable materials, levelled its titanic artillery against the already long ago constructed barriers of the Appalachians, the Hartzgebirge, and the Ural mountains, shattering their crests, and whirling their fragments onward in the midst of its own sandy and gravelly chaos. The first view is admirably described in Sir Charles Lyell's Geology; the second is ably advocated by the glacialists of Switzerland with Agassiz at their head; and the third will no doubt receive all the support and enlargement of which it is capable in the expected book of Prof. H. D. Rogers.

57. It is not in our power yet to say how far the latest **drift** may be a repetition of older events—how far it will explain,

where explained itself, the conglomerates beneath the Potsdam sandstone, the conglomerates of the silurian and carboniferous eras, and the coarse deposits of still later times. But there seems no good reason to confine this sort of action to a date immediately preceding the appearance of man upon the earth, seeing that upon either of the advocated theories we call in no new kind of power, no uncommon forces of nature to produce the drift. Icebergs, nevé ocean currents and earthquake waves have no doubt been ready in all ages to repeat their work. The prime point of interest lies underneath them all.

58. The **plication of the crust** of the earth must be explained. Two irreconcilable hypotheses are advanced. Eliè de Beaumont and his school propose the gradual cooling of the earth-ball, the consequent shrinking of the hardened crust, the sideway thrust of the rocks, and of course their folding up or fracture. Rogers, on the contrary, proposes earthquakes of unheard of magnitude and violence, originating in the same shrinkage of the crust, and caused, immediately by its fracture and collapse upon the surface of the central nucleus of lava. These earthquake lava waves he thinks advanced two ways from certain axle-lines of rupture with irresistible velocity, and in their passage waved the crust which floated like a pellicle of collodion on a glass of water, like a carpet up and down; in fact, we know that every earthquake now does this, but in an infinitely less degree. These waves of the crust became, he says, in some way fixed. Cracked in long lines parallel to their axes, wedges of lava would be pumped and hardened into them, permanently supporting their arches and depressing their basins. While this was going on below, the Arctic Ocean was rushing across the continent above, cutting off the summits of the arches down as far as and below the level of the basins.

59. This theory, whatever may be its faults, is so grand a poem and so nearly arrives at explaining all the principal phenomena of topography, that we may say at least, it should

be true. And none feel for it a profounder admiration than those who collected the materials for its construction. Whether it receive the seal of time or not, it will remain in the history of our science one of those glorious dreams which do more for the intellect of man than verities can do, and therefore justly confer an immortal fame upon the dreamer.

It is easy to advance objections to it; how can a fugitive wave beneath produce fixed waves above? where are the lava dykes which support the arches? and how could the northeast southwest axis of Appalachian force affect the northern ocean and produce a southward or quaquaversal polar deluge? But a single law of form claimed by the Pennsylvania geologists as their discovery, goes further to support this theory of Rogers, than all the objections to it do to overthrow it. It is claimed for the anticlinals of the Appalachian region—a certain NORMAL CURVE—according to which the western rise in front is always long and gentle, and the western plunge behind is always steep and quick. But this is the normal curve of water waves; especially of waves of translation, of breakers advancing upon a shoal or beach. It is true that the arches of the crust according to this theory could have been caused by *waves* of *oscillation* only, for the nucleus lava could have had no shoals, no beach to move on. Yet if the "normal curve" *could be established* in connection with all our anticlinals the argument which it would afford for fluid progression would be irresistible. Here, however, the discussion halts. The author has had unusual opportunities of verifying this "normal curve," if it could be done, and has employed it throughout the illustrations of the Final Report of the Survey of Pennsylvania. But he has found so many exceptions to its rule—so many abrupt eastern and gentle western dips, some of them of magnitude and in most awkward places, and whole *groups* of such reversions of the normal curve, that he feels a serious doubt of its existence. For instance the following Fig., 63, is a careful drawing of the structure of the Chestnut Ridge, one of the best developed of all our great anticlinals in Pennsylvania. Its movement is wholly *eastward.* Its eastern dip for miles is 70° to 80°. The Broad Top region, which contains at least five anticlinals, has all its *steep* dips upon their *eastern* sides. And this instance is the more dangerous for the theory, inasmuch

Fig. 63.

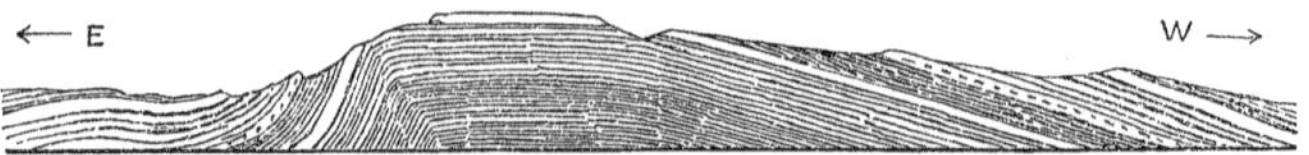

as it rejects the only excuse which the theory has ever offered for such *abnormal* curves, namely, that they are only apparent deviations from the form, in that they occur only when a suite of smaller anticlinals are riding on the back of a greater anticlinal, and therefore when those on the eastern side are tilted eastward, and those on the western westward, so that the west dips of one-half of them are inclined too little and of the rest too much. But here on Broad Top, the *gentle* western dips are *descending the western* side of the great anticlinal of Jack's Mountain, and yet are not so steep as the eastern dips.

60. But more than this: on carefully drawing the curves which the stratification makes across the country on an equal vertical and horizontal scale, it will be found that in a large number of cases—how large a number, whether a majority or a minority, I am convinced that no one knows—there is no such thing as a normal curve at all, in either direction, east or west; but on the contrary, the changes are short and sharp, the flat dips occupying wide spaces and giving place suddenly, and but for comparatively short distances to abrupt steep inclinations. After careful and extensive observation, I feel sure that the forms into which the Appalachian Crust has been contorted are not wave curves, but curves of lateral pressure, involving none of the difficulties of the wave theory, but simply subject to the laws of surface denudation already described. If it be proven, however, hereafter that the "normal curve" exists, I shall joyfully return to the defence of a theory which I was one

of the earliest to accept and illustrate, and which I cannot now regard in any other light than as one of the most brilliant inspirations of scientific genius.

The theory of lateral pressure, however, answers every purpose, provided we imagine the force applied from the south-east (the region of chief disturbance and the exclusive ground of subsequent (?) trap or lava outbursts), and the plication assisted by perpetual vibrations of the crust, not indeed powerful enough to raise the arches instantaneously, but powerful enough to keep it in a plastic state—to shake the strata loose so that they could slip (as we see that they have done) upon each other—and to settle down the arches as they slowly formed into that flattened shape which most of them retain. This may be called the wave theory in disguise, but it is not. It bears no resemblance to that imaginary procession of cosmical billows so grandly painted in Pandemonian colors by Russell Smith, the booming roar of which may well be conceived of as invading the silent regions of the moon, or, perhaps replying to its then resplendent pyrotechny. It is merely what is going on at present in the monthly and daily elevation of the Andes, of Scandinavia, and of other portions of the earth. The cataclysmic part of it relates only to the shaping of the surface finally.

Topography throws itself into a short and simple classification of primary and secondary forms, involving all that has been said before:

1. **Structural Topography**; when the great outlines of the surface conform to the great planes of stratification beneath the surface; as in the Chestnut Ridge (Fig. 63), the ends of anticlinal mountains (Fig. 36), the middle portions of all the coal basins, the plateaus of lava currents, volcanic cones, deltas, prairies, dykes, &c.

2. **Antistructural Topography**; where these two prime elements are in opposition, as in the ends of all synclinal mountains and coal basins, and in cross denudation

generally. Under this head all topographical forms are the result of denudation, and may be subclassified under :

A. **Longitudinal Cutting,** producing the system of parallel crests and valleys and terraces.

B. **Transverse Cutting,** producing cross valleys and gaps, notches, coves, and eddy-hills.

C. **Rectangular Cutting,** combining *A*, and *B*, and producing the peculiar right angle tacking course of the Juniata and other Appalachian drainage waters.

D. **Oblique Cutting,** by which the waters are kept along the outcrop of a soft rock, at the same time descending with the dips, and advancing with the strike. This is a form of the utmost importance in coal operations, and needs a chapter of illustrations to itself. It offers sometimes the greatest facilities, and opposes at others the most onerous disabilities upon the miner, while it crowds the field of the general geologist with problems of the highest beauty and interest. All coal fields exhibit it habitually, but it is uncommonly prevalent and important throughout the Broad Top basins, on account of their gentle western dips and short, steep interruptions. It is finely seen in the Raystown Branch hugging the Terrace Mountain ; in the Standing Stone Creek undermining the slate hills, at the foot of which it flows ; in a word, wherever a stream is seen flowing for miles between long sloping meadows rising into distant hills on the one side, and steeps of slate and cliffs of sandstone on the other.

The Topography of Coal Regions may be classified differently. There are in fact four orders of forms assumed by the topography of our coal districts.

I. The **Horizontal ;** as in the great western basins, and in the centres of all wide, shallow basins ; never perfect, for absolute horizontality is impossible ; the drainage apparently in all directions but really down the dip however gentle, which in many places is but 10 or 20 feet to the mile ; the valleys deeply sunk, with high bluff walls of rock alternately on one side or the other, where the streams strike from side to side through "bottoms;" frequently strange approximations of neighboring valleys, leaving a mere wall between, as between the Mahoning and Red Bank branches of the Alleghany River; or extraordinary horseshoe bends miles in circuit, as at the Viaduct above Johnstown, where the Conemaugh, after a tour of three miles, runs within thirty yards of its upper bed and eighty feet below it. A glance at the map will show the Alleghany River winding along the middle of the broad westernmost coal basin (which has a rise of $\frac{1}{7}^{\circ}$ to the northwest, and a steeper dip on the southeast), down the sides of which its branches flow. The case of the Ohio and its affluents from the west has been already cited. In Virginia, the long arborescent waters of the Kenawha, Guyandot and Sandy rivers flowing northwestwardly to join the Ohio show the gentle slope of the middle of the great Appalachian coal field, although this is interrupted as in Pennsylvania by anticlinal rolls, through which these rivers break, as do the Youghiogheny and Conemaugh waters further north.

The characteristics of this order are the terrace, the hog's-back (or hawk's back) ridge and the horseshoe bend. On the waters of the Kenawha, where the sand-rocks above the lower coal system form the upland, the surface is a mesh of hog's-back ridges, sharp and radiating, with the steepest possible sides and ending in terraced prongs. In the west, these curious remnants of the general plateau are occupied by Indian forts.

II. The **High Synclinal,** as in the Finger Mountains of the terminal basins, described above in p. 79; long and narrow mountains between anticlinal valleys with straight steep sides, and plates of conglomerate on the top supporting patches of the lowest coals, and cleft lengthwise and crosswise by deep valleys.

III. The **Low Synclinal,** as in the bituminous basins nearest the Alleghany Mountain; synclinal valleys ten miles broad, bounded by high and regular anticlinal mountains of the lower rocks—just the reverse of the last order. The Ligonier Valley, between the Chestnut Ridge and Laurel Hill, is a typical example.

Fig. 64.

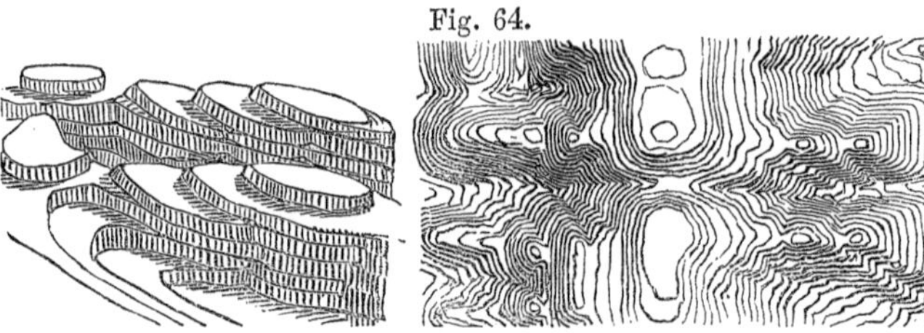

Its section has been given in Fig. 63. Its topography and its theoretical structure are exhibited in Fig. 64, which will repay an attentive examination. The top of the mountain is composed of nearly horizontal fragments of the uppermost member of formation X. The same stratum descends at different angles on each flank, forms the highest terrace, and at times the highest row of knobs. The red shale of XI is next seen, and just outside of it, a row of sharper and lower peaks is formed by the broken outcrop of the great conglomerate. Outside of this again, still lower and lower on the mountain to its foot, on either side, lie the first sandstones, slates and coal beds of the measures. These plates are broken through by the cross denudation and cut down in pyramidal form so that scores of ravines arboresce symmetrically between them. The artist may complain of the stiffness of these forms, but they are true to nature; and nothing is more exciting than to explore this mountain anatomy, thus precisely and mechanically dissected out by the scalpel of a universal erosion. It can be likened to nothing but to that admirable weathering of the surfaces of limestone fossils to which more than to any other cause the palæontologist owes his success in finding them as specimens and his comprehension of their structure afterwards. In fact, every mountain of this order is a gigantic fossil with its interior organization weathered out in a most readable topography; the ponderous hammering of icebergs, and the searching subtle polishing of moving sand and stones, has revealed con amore all the secret tissues of the mass. By its terraces and knobs and their plate edges sloping up and down the several ravines the tracing of the beds from one end of the mountain to the other admits of no mistake. To the distant eye these knobs with their necks of attachment coalesce in perspective and form terraces at different stages up the immense flank and stretching forward far as the eye can see. At the gaps, which are twenty and thirty miles apart, the knobs prevail, elsewhere the terraces. Midway between the gaps high water-sheds fill up the valley to half the height of the mountains and send the drainage waters each way towards the gaps. These waters run in twin streams, each along the foot of its own mountain, leaving a high medial ridge of Barren Measures, on the summit of which occasionally comes in the Pittsburg Coal protected by its limestones. And here we have an adaptation of the second order to the third—the apparition of small Finger Mountains in the bosom of synclinal coal basins.

The manner in which this order of forms suffers a radical change at the far south by the rupture of the anticlinals, and the denudation of the coal on one side, has already been discussed.

IV. The **Collapsed and Compound Synclinal**; as in the anthracite and semi-anthracite basins. This can only be described by reference to what has been said in the preceding pages. It embodies all the forms of the mountain and its valley, its terraces, gaps, eddy hills, oblique and zigzag outcrops, so characteristic on the grander scale of the whole Appalachian topography. The beds descend at all angles, even more than vertical, thrown over on their backs—snapped and the edges slipped past each other—crushed by small rolls running obliquely through the sides of the greater anticlinals, each one of which carries a dozen small ones on its back; their outcrops sweep round sharp points and carry the underground gangways round with them; the surface conforms to all the vagaries of the interior, and betrays them to the observer.

Fig. 65.

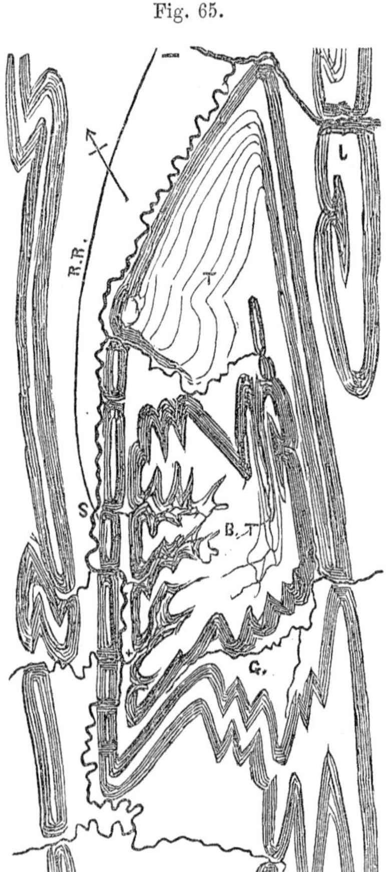

The **Broad Top Region** is an apt instance of this composite order. Fig. 65 gives a sketch of its form and environs. B T is the high plateau of the coal traversed by one great anticlinal which makes the deepest sinus in its northern scarp, and indents most deeply the mountain south of Ground Hog Valley, G, but is lost and flattened away in Trough Creek Valley, T. Other smaller parallel anticlinals assist in breaking up the basin as a whole into five or six subsynclinals, each of which has its own terminal finger to the north. The coal beds sink rapidly from the top of these down to the level of Six Mile Run, the middle of the three drains which the mountain has towards the west, and then rise as rapidly to the top of the southern margin. Shaub's Run, on the north, and Sandy or Castillo's

Run on the south also trench the mountain transversely to its base, exposing the beds in all the little basins. Three branch roads down these valleys meet the Huntingdon and Broad Top Railroad, which runs from Hopewell (X), past Stonerstown (S), to Huntingdon (H), where it delivers its cars upon the Pennsylvania Railroad track, or shutes the coal into canal boats. The mountain of coal is seen to lie in a circular valley of red shale (XI), surrounded by a mountain of X, called Terrace Mountain on the west, Sidelong Hill on the east, and Harmer's Mountain on the south. The Raystown Juniata breaks through this twice, and flows awhile in the red shale. Around the whole sweeps the valley of Upper Silurian, outside of which again runs Tussey Mountain, IV, and the Lower Silurian rocks beyond. L is an anticlinal of IV, on the east, the southern end of Jack's Mountain and the Kishicoquillas Valley.

CHAPTER IV.

TOPOGRAPHY—AS AN ART.

1. WHEN truth is made the beginning, middle, and end of art, art becomes both beautiful and easy. The laws which govern its relationships to science are not conventional nor empirical, but pre-established and perpetually real. Art is the creative self-consciousness of God, the Logos of Nature, the utterance of science. What is not a divine thought cannot become fact; what does not exist cannot be pictured; what is untrue cannot be beautiful, but must involve the eye and the hand in obscurity and confusion. To study nature, therefore, with a perfect love for truth, to pursue science with inextinguishable ardor and ceaseless devotion, is the only valid claim a man may make for himself upon that exhaustless treasury of powers and beauties, over which preside the Genii

of Expression. Invested with the nobility of a genuine extraction from the eternally true and real, every group of facts in science presents itself to art as worthy to be said or drawn; nothing real is ignoble; the terrestrial is celestial; the face of the earth is the face of a great angel, with infinite smiles and anguish lines and profound sympathies with peace and suffering stamped upon its features. Every lineament is a line of tragical history, full of pathos and sublimity. The Topographer, if a true artist, will put himself in true relations with this grand mute object of his study, and learn its own record of its wonderful experience, if he will picture the earth as it is. The draughtsman must be first a geologist.

2. As every movement in the human body reveals and is in turn revealed by some muscle, and every feature of the face is characterized by some passion and finally fixed by some inveterate habit, so every trait of the earth's surface is indicative not only primarily of some rock structure immediately beneath, but also of some past elemental, long continued, frequently repeated, localized, and graduated action of its forces. The ridges on the earth show as much the practical energy of ideas as the lines on the page of a book. Had a wilful fancy moulded the earth this would not be. But, what else *could* result from its gradual and regular formation under the direction of various but constant and well balanced vital laws of force and motion? To learn to read the surface aright, is therefore the first lesson in geology. To map the surface of the earth is the indispensable condition of the progress of discovery, and will be the crowning achievement of a perfect knowledge of its geology.

3. As the face of youth is difficult to paint from the swift number of its passing passions, and the face of age is difficult to paint from the multitudinous modifications of the original features by the long series of events they have lived through—and yet the prime elements of the human countenance can be counted on the fingers—so the face of the earth, simple in its constitution, and simple in the primeval laws under which that constitution has continued to exist, is nevertheless so crowded

now with vestiges of the past, so scarred and pencilled over with the records of innumerable adventures, that if anything is true of it, it is **Detail.** This is the characteristic distinction, therefore, also between false and true art. As government grows vicious *pari passu* with centralization, in which the detailed forces of society acting in individual liberty are swathed and palsied, frozen up and turned to stone, by an unnatural process of solidifying known only in the artificial world of politics; so the whole liberty of the natural mountain vanishes from it when the false artist paints it on his canvas as a single mass, reducing what is a world of facts to one bald falsehood, or ennobling a small group of falsehoods at the expense of many worlds of facts. The noble public protestations of Ruskin against this crime habitual to false art, have been whispered to himself by every intelligent traveller, who ever stood upon the Righi, or the Dom Platz at Berne, who ever feasted with hungry eyes upon the marvellous infinity of cusps of Mount Pelvouz, or drank the interminable flood of lines poured from the Maladetta along the summit of the Pyrenees. Upon the map as well as on the canvas **detail** is truth, is everything. The foliage of a tree is not more multitudinous than the pencilling of a cliff; the history of an empire is not more complex or more difficult to reduce to system, than the surface of its territory.

4. Yet as has been said, and I hope shown in the preceding chapter, nothing can be simpler than the rules of construction and denudation which govern all topography; the difficulty lies in the detail; and no part of an artist's duty is easier than to learn and to apply them; whether he be a local railroad draughtsman, or a professional map-maker, or a landscape painter, accustomed to the absurd piles of incomprehensible material called "rocks," and the fatty impossibilities of distant "mountains" on the canvas of the great masters. Practical directions for map manipulation will be given shortly, but the author here desires only to insist on what will now be universally granted, but should come practically home to the heart of every

artist, that the artist has no excuse who pretends to use his eyes and study nature for not repudiating the unnatural and impossible from his representations. It may be all very fine for Martin to project table-rocks as large as Jupiter's satellites from the summits of Mount Blanc, when he paints the Creation, or the Dissolution, or the Deluge, or Pandemonium, but it is not lawful for ordinary artists painting the common aspects of nature to ignore the fact that things can't stand upright unless they be supported; that there are definite limits to the strength of granite; that all rocks show *regular* lines and bands, either of deposition, or of crystallization; that the *roches moutonées* of New England and of Scandinavia do *not* look like coprolites, nor like hardened lumps of green and red putty. These are universal blunders, and yet too obvious to be explained, except by taking it for granted that the artist paints in town, and copies lithographs in oil.

5. But there are other things to know. It is not only needful to be able to paint *bark* upon a tree, but to be able to portray the law of posture common to a *grove*. Just so in drawing rocks, it must next be learned that all the rocks that enter into one picture bear a scientific relation to each other —all the hills in a vista—and all the mountains in a background; and the background cannot be one thing by itself and the foreground another by itself. Every range of sedimentary rock cliffs *must* exhibit certain lines which class them in the picture as a whole and intimate their common structure. We have seen on a canvas the strata in the two opposite walls of rock inclosing a cataract *dipping different ways!* A phenomenon so extremely rare in nature, that no artist may pretend that he has been the man to see it. It is common to observe in pictures granite knobs projecting from sandstone escarpments, like the dismayed face of a village orator from the hogshead into which he had broken through—or primary aiguilles shooting into the air from the summit of bossy decomposed lava slopes like ship's masts stuck upright in a vegetable garden. It is even not enough to be in a condition to avoid

these errors ; one must know still further, and can easily know certain niceties of relationship between the Within and the Without, the formation and the surface, and certain mechanical methods which nature has, not sparingly and locally, but universally and consistently adopted for producing her exquisite delights for the eye. The dash of the ancient current down a side valley will explain a grand circus of rocks on the opposite river shore. The climbing of a massive central diaphragm of flint to the summit of a narrow ridge, the sharp outline of which cuts against the sky, should appeal in a picture to the intelligence as well as to the fancy ; it does so in nature.

6. Practical topography, however, is more difficult than the simplicity of its rules will lead us to suppose. The common methods by which a sketch of general resemblance to the ground is offered in lieu of a true portrait are most mischievous, because they not only fail to present the actual facts upon which alone correct reasoning can go on, but they distort and even reverse, so as to *oblige* deductions to be wrong. For the sake of filling up vacant spaces they suppress indispensable facts and substitute for them pure inventions and thus they misguide. The same reproach lies against the common geological profiles, which are vertical sections of the ground, intended to show the relief of the surface and the structure of the crust to a certain depth. These are commonly mere distortions and caricatures which illustrate the subject by confusing the observer. Neither maps nor sections are helps to the geologist—and are cruelties to the schoolboy—unless they be properly and accurately made. Then they become his efficient tools. But he must make his own tools. He will find none made to his hand. The age of maps has just set in. Even yet maps are made from horseback or railroad embankments ; beyond a hundred yards each way, are mere guesswork to fill up the sheet. Surveyors' charts are still notoriously bad, show no topography, and even in their first elements are self-conflicting and unreliable. A normal school and board of control are absolute necessities of the profession. No one, in

fact, seems to know what a perfect map should be; no one wants one; and yet what is a geologist—*what is his geology* without one? That none but a geologist can make a map is evidently true from the fact that we only see what we look for, and the geologist alone looks for *surface indications of internal structure;* he knows, therefore, the importance and significance of what to any other man is nothing, or at best a curiosity. To him, the landscape is alive with more than vegetable and mineral wealth; he sees in it an informing soul—life pulsating in the vales, and heaving in the hills. He feels more than the warm or cool winds, hears deeper sounds than the low of kine, the rumble of wains and the murmur of sequestered cataracts. The Old Forces are to him only sleeping, perhaps dreaming of what they have once done, perhaps pondering their next developments. He gazes at a country as Vandyke or Raphael divined a noble face, and painted its whole history in one look. A true and perfect map must be a history of the geology of its region, and should tell its story uncolored. A pale face may be as impassioned as a florid one. The mechanical process of making such a map will now be described.

7. **Map-making** divides into Field-work and Office-work. By the first we collect our materials, and by the latter realize our ideas. Field-work divides itself into Reconnoissance and Instrumentation; Office-work into Calculation and Drawing. When the ground has been first reconnoitred to discover its extent, its general features, and the best method of attacking it, it is then measured and levelled in determinate lines, the calculations of horizontal and vertical angles follow, and the projection of these relations upon paper is the Map. In the field, therefore, there are—Instruments to handle, Note-books to keep, Roads, Streams, Cross Sections, and Outcrops to run, Levelling by spirit-level, by vertical circle, by barometer; and in the office—Notes to be reduced, Lines plotted, the Map covered with its real and ideal symbols, and Cross Sections

constructed to exhibit the geology. I will pursue this order in the few pages that the limits of a Manual still permit.

8. **A. FIELD-WORK. a. Reconnoissance.** This ought never to be neglected. Hurry is no excuse. More time is sure to be lost by adopting wrong lines of impromptu instrumentation than can be spent in the most careful reconnoissance. The geologist should go over the ground on horseback first, with some near resident, to get the "lay of the land," the run of the roads and paths, and the salient features of the topography; and then on foot, to establish starting-points, and recognize land lines or corners, cliffs and springs. A definite plan of field-work can then be intelligently devised, and time and energy saved in its execution. It often happens that for want of a reconnoissance the key facts of a survey are discovered only at its close; lines are found to have been run backward, or so as just to miss what they were intended to reveal; and thus whole masses of material turn out to be almost worthless.

9. **b. Instrumentation.** For reconnoissance there is in use, of course, the pocket-compass and the pocket-level, the aneroid strapped upon the breast, the foot-rule with a spirit level on one side and a graduated arch between its legs to use as a slope instrument, a half-pound hammer, a good lens, a bottle of acid to distinguish iron from lime, a few wax tapers for deserted gangways, a satchel and a note-book. For work, a party of five or seven men is needed—a compass or transit, a 50 foot chain, and pins, hatchets to clear the brush and make stakes, axes where the woods are thick, a level and target, and some kiel to mark the stakes and trees in red. Most of these things are in constant use and need no description. A transit instrument, however, is too heavy for ordinary topographical work.

10. I would recommend, therefore, as the best form of **compass,** one mounted with a ball-and-socket joint on a light tall tripod, with four side adjustment screws, and two small levels at right angles to each other on the face, one in front and one to the left; to the right is attached a transit telescope playing

vertically against an upright circle, graduated to half degrees, and read off by an index, on each side of the index line of which is another line the fourth of a degree distant. The use of this will be described under the head of levelling. As the axis of the telescope is out of line about six inches to the right, a lopsided 12 inch target must be used instead of a sight-rod, the sliding disk of which divided into four alternate red and yellow quadrants has its centre point at an equal distance to the right of its staff with the eccentricity of the telescope from its tripod.

11. **Odometer.** All preliminary work should be done with an odometer with a ten foot wheel, made like the patent buggy wheels, tired with copper plate $\frac{3}{8}$ inch thick, bolted through the felloes. The gas-pipe wheels in use will not stand long on rocky roads and are sure to break down in the places where they give the operator most inconvenience. The body of the odometer should be made of the stiffest and lightest wood and braced with copper wire; as also the Jacob staff behind. If the indices read double ten double thousand and double ten thousand feet, the note-books must have a final blank column on the page to allow of doubling. The odometer can be used even in the woods. I have thoroughly tested it on the roughest and most irregular ground and can assure topographers that year in and year out it is much more reliable and will do much better work under all conditions of survey than the best pair of chain-men he can train. The errors of the chain it is almost impossible to help or to discover; they are irregular and affect a whole line. The errors of a good odometer are slight and of great regularity and subject to compensation laws, or are errors of single readings which usually contain the elements of their own correction, but if not, affect only two distances in a line. In cases of extreme necessity the author has surveyed a tract *alone* with odometer and aneroid, taking front and back sights, doing the work of a full party of seven, and with entirely respectable results. No deduction need be made for the difference between hypothenuse and base, hill slope and dead level

measurement, on common roads, for these never exceed 6°. But on cross sections a table of natural signs must be used which will deduct **one foot per hundred** for a slope measure of 8°.

Table of Deduction for Slope Measurement.

.01	.02	.03	.04	.05	.06	.07	.08	.09	.10
8°	11½°	14°	16¼°	18¼°	20°	21½°	23°	24½°	25¾°
.11	.12	.13	.14	.15	.16	.17	.18	.19	.20
27°	28¼°	29½°	30¾°	31¾°	33°	34°	35°	36°	37°

12. c. **Note-Books.** The best form of note-book is a great desideratum. I have long used Hufty's plotting sheets bound up, employing one page for sketching and the opposite page for notes. All notes commence at the *bottom* of the page and ascend. All stations of the compass are numbers *in rings*, which distinguishes them from all other numbers. These occupy the first column, the odometer readings the second, the courses the third, the angle of elevation or depression the fourth. The fifth is left blank for the actual distance (twice the difference of the odometer readings), and the sixth for the actual difference of height (when calculated) between the present station at the compass, and the next station at the target. All these should have drawn round them *cartouches*. All side observations from a station follow above it, plain, on the page. All measurements and notes along the interval do the same. No attention ought to be paid to saving paper; it is the worst economy; small note-books are a nuisance. Everything should be sacrificed to an open, clear, roomy page, crowded with sketches of houses, station trees, fence corners at a distance, hill-tops, &c. The opposite page should receive a running sketch map in contour lines drawn from the bottom of the page upward in the direction of the work and of its notes.

13. d. **Road Running.** This should be preliminary and very accurate; therefore with the odometer. It is bad to mix odometer and chain work, for however they may agree on a trial, they work with a difference, and a variable difference,

because chain men are frequently changed, and even the same men measure differently when trained or untrained, fresh or tired, sick or well. But the wheel works always alike. The road running should proceed not in straggling lines radiating from head-quarters, but in circles, which close and afford opportunity for proving the work to be correct. Time, however precious it may be, should be spent in plotting up this preliminary work as fast as it proceeds, so that all errors may be at once detected and corrected, and an unchangeable base be obtained for the topography. When a survey becomes extensive and complicated with lines, unless the primary polygons are well established the data collected become a hazy nebulous mass, doubtful everywhere and causing untold anxiety, labor and time to get into order. Hence all *first* work should be slow, exact and well worked up in the office as it goes. Afterwards the notes of cross lines may accumulate safely. Road running finds its chief use in this : its distances need no deduction for slope, and it sketches out the ground for a full survey.

14. In running roads the centre should be kept ; stakes set on the right side running, facing the road, and marked from 0, up to 10,000, on the same survey. No two stakes in the same district should bear the same number. At cross roads, pegs should be driven flush with the surface, and large stakes set to the right. All fences coming down to the road right and left, and all fence corners of fields right and left should be measured as passed, and marked in the notes F R, F L, F C R, F C L. All creeks, spring runs and drains should be marked when crossed; for it is never possible to foresee where a new line may be taken up. All house and barn corners should be likewise noted when passed, and the distance off the line paced in.

15. Stations should be set if possible at the end of every long fence, so that its course may be taken and then the bearings of its several high and low places and distant end from the next (or previous) station will give the best means of sketching in the topography. Every conical hill-top should be

split, or its tallest tree-top selected, and its bearing noted from several stations. The *first* observation of a distant point stands merely after the station, &c. from which it is taken; when taken *again*, the note is in *brackets* preceded by the number of its station; when taken a *third* time (and three observations should always be taken if possible) the note is in *double brackets*, with the number of the new station, &c. When this is once done, any number of points can be triangulated to, and the fact of their being unbracketed, bracketed, double bracketed, &c., will of itself identify the different bearings with the different stations from which they were taken.

16. e. **Stream Running** is only necessary in the case of large waters, but wherever it can be done it does more for the finish of the map than anything else, and is of great service to the geologist, leading him surely to the best exposures.

17. f. **Cross Section Running** is of the utmost utility. When a country has been made out by its roads and the general features of its geology determined, especially the strike of its measures, straight lines should be surveyed across them, beginning at well marked stations on the roads, staked every one or two hundred feet, and slightly accommodated to the accidents of the surface. The object of these lines is first to give the shape of the ground and therefore stations should be made at every marked change of slope; then thereby to discover the outcrops of the rocks on the sides of the hills. When therefore such a line happens to be laid down or up a cross valley, it is well to make an offset at right angles large enough to shift it on to the hilltop again. The cross sections will of course be parallel, and may be numbered A B C, and their stakes A 1, A 2, B 1, B 2, &c. But it is much better to include them in the series of lines, and let their stations fall into the whole series of numbers.

18. g. **Outcrop Running** is the complement of cross sectioning. It takes up the cliffs, the terraces, the streaks of ore fragments, limestone crops, coal crops, &c., at the cross section lines, or at some road station or other established point,

and follows them *along the strike,* making branch lines down over the terraces, &c., every few hundred yards, so as to watch the effect of the geology upon the topography, and to detect those changes in the sandstones and other key rocks, which give so much trouble unless well understood. If the geology be difficult, outcrop running should be begun early in the survey and be repeated at intervals so as to clear up questions, and prevent the accumulation of conjectural theories.

19. **h. Levelling with the Spirit Level** should be adopted upon the base lines of a survey. Two good men can level from 4 to 5 miles of *mountain* road lines per day. It is unsafe to trust the vertical circle for a complicated survey, because its compositions are so numerous and difficult as to overwhelm the office worker. The spirit level runs rapidly over the main lines, keeps all clean and straight, and leaves the vertical circle compensations on cross lines simple and easy.

20. **i. Levelling with Vertical Circle.** This is done by having a long spirit level mounted on the telescope tube, with a hole in the index and a brass plug, stopping it at zero ; then level the telescope with the adjusting screws, slip out the plug, sight up or down at the target, and note the angle. With three lines on the index $\frac{1}{4}^{\circ}$ apart, observations to the 5′ can be accurately and speedily made, and practice will enable one to read still smaller divisions. When the tripod is used the target must be slipped up or down to a level with the axis of the telescope at every setting, and before the target man goes forward to the next station. If the odometer be used this is unnecessary, except upon very uneven ground. Sighting at fence corners aside from the line, it will be near enough to sight at the top rail, but where the target man makes measurements by pacing his target goes with him and will give the height. The accuracy of vertical circle levelling is very remarkable, and can be explained only by the laws of compensation which all repeated action obeys. In running a spirit level after a vertical circle over the same line, the data of the latter will be found to lie above and below those of the

former alternately, involving a series of errors, each one local to the stake, but the whole series so self-compensating as to leave a final error of usually but a few inches, or at most a few feet in many miles. I have repeatedly levelled thus a polygon of eight or ten miles, and come out within a foot or two. For all the purposes of geological topography the vertical circle which levels a line as it is run is as good as the spirit level, while it dispenses with a leveller and rod man, brings the level notes under the same eye and hand which is discussing bearings and distances, and is available for distant objects, which the spirit level is not.

21. j. **Levelling by Barometer.** The mercurial and boiling point barometers have been given up by ordinary topography. The Aneroid is under discussion. The author has had a large experience with it and gives to it his unqualified approbation and affection. He finds it hard to see how a geologist can do without one. He has made extensive surveys with Aneroids alone, and although both field-work and office-work are excessively laborious, they are equally successful when the aneroid is carefully handled. The rules are few. No observations more than five minutes apart are to be compared, without repetition; no observation is to be made at a station until time has been allowed the instrument to come to rest, especially when an ascent is changed to a descent, and *vice versa;* all observations followed by thunder-storms are to be of subordinate value; all lines of level are to pass as nearly as possible across from one base line to another, parallel with it. These rules observed, the results of aneroid practice with the best and latest instruments are surprisingly good; for practical topography, perfect. Each instrument however must be experimented upon before being taken into the field with ice and sunshine to get its scale of thermometric compensation, for the tables accompanying the barometer are worthless. Some instruments are nearly, and a few quite, self compensating within ordinary working limits, say two thousand feet; but others require an allowance of five feet to every degree the

thermometer rises or falls. The observer must of course learn to know his instrument well, or he can do nothing with it.

22. **B. OFFICE-WORK. Note Reduction.** This should be assiduously kept up abreast of the field-work. It is but too apt to fall into arrears ; if it do so, the observer is in danger of bankruptcy, overwhelmed with fictitious wealth, the over-expanded currency of science. Imagining himself accumulating a fortune of facts, he is but rolling up before him work which will crush him in the end. It is a sadly common error in observers to gorge themselves with undigested and finally indigestible facts ; to spend a lifetime in becoming intellectual stomachs, armed about the mouth like octopods with long tentaculi, which grope on all sides everlastingly to take but never give. There are representative men whose names stand high in science as appropriators, not as benefactors ; who form a cuttlefish class among their fellows; but of these I do not speak. But there is many an assiduous *worker* who works too long, too hard, collects too much, rests and reflects too little, employs the compass where he should employ the pencil or the pen, and sinks at last into an unhonored lethargy, drugged with the opium of his own poppies, ruined by the number and fulness of his own note-books. It is a well established fact that a man after all must do his own work however many he may get to help him, and the larger his parties and the more zealously they help him, the more he has himself to do to lead them, think for them, and bring forth order out of their confusion.

23. Every night the blank columns of distance and elevation should be filled up on the pages of the day's work, the one in red ink, the other in blue. N.B. No lead pencil should ever be allowed to touch a note-book page except upon the ground; all after corrections, additions, calculations, suggestions, explanations, should go down in red ink, so as to leave the original record one of safe reference. For the same reason all corrections of a cipher on the ground should be repeated over it immediately to relieve the office-worker of all doubt.

24. The **reduction of odometer distances** is easy. Begin at the bottom of the page, subtract the first odometer reading from the second and multiply it by two. To check the page to see if it be correct, subtract the first reading at its foot from the first reading at the foot of the next page and multiply by two, the product ought to equal the sum of the distances on the page.

The **reduction of vertical circle angles** requires either the logarithmic tables of the Nautical Almanac office at Cambridge, Mass., or the table of natural signs. Odd distances are best calculated by the former—even hundreds and quarter-hundreds by the latter. For example 247 feet, up 7°15′ would be calculated thus:—

247, log. 2.3927
7° 15′, log. 9.1011
1.4938 log. = 31.16 feet.

While 250 feet would be calculated thus :—

7° 15′ nat. sin. .1262
$.1262 \times 250 = 126.2 \div 4 = 31.55$ feet.

In all instances some convenient upward and downward dash must be affixed to the angle noted to express whether it be up or down.

25. **C. Plotting.** Good plotting is the result of long practice, a temperate appetite, a good eye, a steady hand, well made instruments, and entire absorption in the work in hand. Conversation in the office is death to fine work; it ought never to be allowed. Nor should plotting continue uninterruptedly more than two hours at any one time; there are despotic limitations to the faculties of brain and nerve. Nor should the work be carried up to the dinner hour nor recommence immediately after it. In the inevitable errors thus produced more time is lost than saved.

26. The instruments of the table should be of the best. The old forms of protracting scale, quadrant, semicircle, paper circle, and circle with movable arm (all of which work from a

perpetually shifting north line difficult to keep in the true meridian) are now all superseded by the stationary Cleaver's Protractor made by Young of Philadelphia. This is fastened with spring thumb-pins to the table, and the paper with its north line adjusted to it. The movable square around the graduated circle of the Protractor can then be made to present one or other of its faces in parallelism with or at right angles to the bearing to be plotted; in the first case a parallel ruler is rolled from it to the station and the course is drawn; in the other case the parallel ruler is rolled against a glass triangle. Two glass triangles and glass straight edge may be used instead of a parallel ruler, and should be if the ruler be not true in its rolling. A prime requirement for good plotting is a smooth flat table, cleated to keep it from warping, and guarded on its front edge to let the paper slip down before the breast of the draughtsman under the guard without being crushed.

27. The **scale** is a difficult instrument to manage. All scales are defective in their first small division, which is made out of proportion to allow for *stickage,* that is for the aberration produced by the points of the dividers descending into the wood. This has been done to accommodate country surveyors and careless plotters. An ambitious draughtsman wants no such mischievous assistance. His dividers never more than penetrate the paper. Whenever he has a few distances in nearly a straight line he tries his measurement with the scale. There is also a small wheel to be had with an index point to test the measurements along a whole line to find any error. It will discover an error in a few moments if any has been made. The best scales are paper scales made by Mr. R. S. Perkins on a machine of his own invention at the U. S. Arsenal works at Frankford. They are superior to copper, steel or ivory scales, because more exact at the end, and fresh ones can be had to replace the worn; they cost about six cents apiece.

28. The author finds his primary **plotting scale** to settle upon 400 feet to the inch. He used 500 for some years, but gave it up as too small to express all that ought to go down

on cross section lines staked every hundred feet. His scale for geological surveys of eight or ten miles square, is 1000 feet to the inch. Geographical surveys involving topographical and geological features, may be made comfortably on 2000 feet to the inch. It would sound much better to adopt some element of the earth's form as a unit of size, a degree of the equator, for instance; or to adopt some aliquot part of the thing itself surveyed, and represent all roads, house plans, streams, hills, townships, one ten thousandth, or one hundred thousandth of their actual size; but our standard foot and inch are fixed, the mile also is an unchangeable fact, while the slight amount of idealism gained by the change would hardly compensate us for its practical inconvenience in the work.

29. **D. The Map.** Its elements are first, the representation of *real objects*, *a*, roads, houses, barns, villages, streams, bridges, springs, cliffs, coal openings, railroads, fences, and the distinction of wild and cultivated ground, and the hillsides and summits generally. Secondly, the representation of *ideal objects with real references; b*, lines of survey and triangulation with their stakes and benches; *c*, lines of property with their trees and stones; *d*, lines of government with similar termini. Thirdly, the representation of *pure relationships*, by conventional symbols; *e*, lines of latitude and longitude; *f*, lines of magnetic force, &c.; *g*, lines of contour or water level, referred to tide; *h*, lines of internal structure, dip and strike.

30. **a.** Roads are represented by parallel lines, drawn with a parallel pen; it should be lightly and delicately done, and the width of the travelled part of the road be kept to scale, for this permits the fences to be laid in zigzag lines, which will be found to be a matter of importance when the topography is to be laid in. Houses, *approximately* or conjecturally located should be represented by a bare parallelogram; but every house *actually* located by a parallelogram, half-filled in with black to show the way the roof runs; this trifle also will be found of importance when large surveys are to be

verified by distant cross-sights. Fences, the course of which is only guessed, should be dotted in; and streams also beyond the point at which positive observation stops. All waters should be laid in on a map—at least on the primary sheets, in blue, to keep them out of the way of the contour lines.

31. **b. All lines of survey,** in a word, all ideal lines should go down upon the map in red ink. The distinction between the real and the ideal, the known and the suspected, should be vigilantly observed. What is a map good for unless it tells its own story? But along lines of survey exist real objects which enter into its composition, and these should go down in black. All stakes should be small black circles, numbered *in red, upon the right hand advancing,* and with their levels above tide (or other datum point) *in blue, upon the left hand advancing.* The importance of this arrangement being carefully preserved will be felt when the contour lines come to be laid in: side trees should be solid black circles.

32. **c. All land corners,** whether posts, trees, or stone-piles, should have a larger black circle drawn around the small black circle which represents them as stations in the survey. An *actual blazed line,* inasmuch as it is made up of real trees, should be a black dotted line upon the map, but the *land line of the deed* should go down also, but in red; the two may, or may not coincide, or even correspond, but it is important to show which is purely theoretical and which has been realized, even if erroneously.

33. **d.** Lines of government, county and township lines, are comparatively so rare that the taste of the draughtsman need be his only guide.

34. **e. Lines of latitude and longitude** are seldom inserted upon American surveys, because it is usually impossible to obtain them with even tolerable correctness. An attempt to compile a map of the United States will prove to any draughtsman's bitter satisfaction, into what a shameful condition our interior astronomical geography has fallen. It is to be hoped that the different States will soon appropriate suffi-

cient funds for determining, by telescope and telegraph, the latitude and longitude of their county towns.

35. f. *Magnetic Variation.*—On every map this should accompany the arrow showing the true north. There is the greatest confusion prevailing among the deed drafts of the various State land offices and the road records of the County offices. The variation of the needle from true north having advanced eastward up to about the year 1820, and then having been retrograding ever since, at a variable rate for different places, while its whole quantity at any one time, is different for every different place east or west of a given magnetic meridian, it is impossible to bring into even tolerable agreement, the courses of any old land line upon its different earlier and later surveys. The law of some States that every county seat should have a variation standard mark, and that all county surveyors should run by true north, is not as well observed as it might be. Few deed drafts say whether they are run by magnetic or true north, and even careful surveyors permit themselves sometimes to be governed by circumstances in their choice. It is a most perplexing and extensive subject, well worthy the attention of a national executive commission, empowered to experiment, investigate, and determine standard points and lines, draw up tables, and publish explanations and instructions. An immense mass of materials are now collected which only need digestion and publication in a cheap practical form to relieve the land surveying of the country from many of its evils. In compiling the geological map of Pennsylvania for the State Survey, the author found certain counties interfere along their north and south lines more than half a mile, and the accumulation of such errors upon the West Branch of the Susquehanna had they not been redistributed, would have amounted to at least five miles in an east and west direction.

It must be remembered that needles are hung differently; are charged differently; read differently at the two ends; that the variation varies with the daily course of the sun, and with the season of the year. Two plottings—the work of the same

18

eye, compass, hand, and scale, but one made in the spring and the other in the fall, may differ half a degree all round. Every map therefore should have its full date upon its face.

The *daily* variation is variously stated at from 5′ to 8′ in winter, and from 11′ to 13′ in summer. It is 0′ about 5 o'clock A. M., moves eastward till near 8 o'clock, is 0′ again about 11 o'clock, and continues to move westward until 1½ P. M. At 5 o'clock it is again 0′, and moves eastward in the evening.

The *actual* variation must be stated thus :—

5 A. M., normal ; diminishes to 7½ A. M., then increases to
11 A. M., normal ; increases to 1½ P. M., then diminishes to
5 P. M., normal ; diminishes until towards midnight.

36. g. **Hachure lines** are to be rejected from our art in almost every instance. They are those radiating, straight or quivering lines by which mountains have almost always been represented. But they really represent nothing in nature. The descending ravines must themselves be represented by hachures, and therefore are not their origin. We see no such lines in the landscape. They tell nothing but the direction and strength of slope, and even these elements in the most imperfect and variable manner. It is impossible to curve the hachures nicely enough to preserve the true line of greatest slope, or to observe their relative distances and strengths according to the purely empirical standard. They perpetually interfere with one another, must be broken and interpolated around all curves, and cover the paper with unmeaning and expensive lines to the exclusion of important data. The only case in which they may be used is in *continental* geography, where the topography is alpine and unknown ; here the sketchy, indefinite, non-committal hachures may be considered as in place ; but even here we usually see all their worst faults exaggerated, and the reins given to their inherent lawlessness. The result is, as everybody knows, a hateful caterpillar-like topography, unnatural ranges of supposititious mountains, crawling hither and thither over the paper, not only teaching no-

thing, but suggesting falsehoods at every step. A few delicate contour lines would not only avoid all this, but would show what is actually known and what it is important to know, leaving the rest undisturbed to the imagination. If hachures are used at all they ought to be used with the utmost delicacy, and vanish in all directions indefinitely into the uncovered spaces supposed to be valleys or lowlands.

37. **h. Contour Lines**, are actually representatives of certain lines in nature; for we see the horizontal line everywhere reproduced, in the crest line, the terrace line, the outcrop line, the foliage line, the water line. But contour lines are representatives of the positive rather than of the actual; of relationships, rather than realities; of what might be, not of what is; but suggestive of what is. Suppose the ocean to rise along all its coasts, and up all its bays, ten feet—or a hundred—or a thousand feet—how different would be the lines its edge would then describe upon the map! These supposable lines are our contour lines; they describe themselves through all such points of the surface as are at the same given elevation above the sea. If the sea be supposed to rise ten feet at a time, and high enough to cover the highest mountains, it is plain that the contour lines so described by its margin would form a perfect picture of the surface of the earth, and moreover announce with the greatest distinctness and accuracy all its relationships of elevation. The advantage of this is open to the most inexperienced eye. A map which brings every point upon its surface to its actual level, must show perfectly the steepness of hills, the outspread of plains, the localities, forms, and directions of cliffs and terraces, and the variable grades of stream beds. Whatever the geologist, the miner, the railroad engineer, or the proprietor can ask for as preliminary to construction or exploitation, a perfect map in contour lines cannot but afford. The geologist will trace upon it his outcrops, and calculate by them the geometric curves of his basins; the miner will select upon it the most favorable and

least expensive places at which to commence his openings; the engineer can make his primary locations upon it without going into the field, and the property holder can intelligently make upon it exchanges with his neighbors, and square up his limits with profit to all parties in the working of their mines.

38. The various niceties in the drawing of contour lines can only be learned by faithful observation in the field, practice in the office, and experiments with solid bodies of various forms immersed in water. Cones, cubes, pyramids, cylinders, fluted and beaded surfaces and finally amorphous masses should be carefully immersed at fixed intervals of depth, and the margin water lines they present copied upon paper. This will accustom the eye to the continuity and separation of the lines, and exhibit a multitude of beautiful relationships of vertical heights to horizontal extent. Then in the field, dead level lines should occasionally be run with the compass, target and chain at short sights, to get all the inequalities of surface, and establish a guide line, to which the cross section line and water line running notes must be obliged to conform. At the same time the side slopes both up and down should be measured, by bringing the telescope sideways round and slanting it parallel with the target rod laid flat at a little distance ahead upon the ground; or by taking right-angled offset sights at the target itself, say fifty feet off, up or down the slope. It is good practice in running upon steep slopes to turn the instrument and sight up or down at the bolls of trees, at about the height of a man's breast; this will train the eye to a very accurate estimation of the strength of slopes. In sketching, the contour lines must be put in as they cross the run line, forward right or forward left, and be made close or distant according to the strength or weakness of the slope. It is easy enough to paint a terrace just where it is crossed, and therefore theoretically horizontal, by observing the due distances between the contour lines; but it requires long practice and numerous observations to paint *a descending terrace*, which, while it is one of the most common, is one of the most im-

portant geological features in the topography, and one of the most delicate to handle with the pencil. Successive ribs of rock descending at a steep slant to water level, converts all the contour lines into zigzags of a peculiar variable shape, perfectly expressive of the geology; and it often happens that what the geologist toils over in vain and gives up in the absence of underground working as an insoluble problem, opens all its mysteries to the first glance at a well made contour line map.

39. **Coloring.**—Contour lines should be laid in *blue*, according to the analogy of the laws of color adopted in map-making; but it is better to draw them in some half, or neutral tint, or some light sepia. Very successful maps have been lately made (in hachures), with blue for the waters, black for the roads, cities, and names, and sepia for the topography; the colors harmonize without obscuring each other. A name can then go down over both a water line and the densest slope lining. So can a red survey line, or a black coal-crop streak, or a crimson trap outcrop, or any other positive rock coloring. Even properties can be washed in over all in broad half-tints.

40. **Relief** can be obtained in a far better way than by hachures, by drawing the contour lines strong upon the shady side of a hill, weak on the two flanks, and alternately omitted on the sunny or full light side. In doing this, artistic taste and practical experiment make their own rules.

41. i. **Sections** are intended to represent the crust of the earth to a certain depth, as it would appear if cleft vertically at right angles to the strike of the rocks. When badly made they are worse than worthless. The best books on geology are full of absurd and mischievous sections. Papers, in the proceedings of societies, which ought to do honor to their authors, too often subject them to the gravest suspicions of not knowing what they are describing, of having misinterpreted their best exposures, or of distorting facts to support a theory, simply because of the exaggeration and distortion of their diagrams and sections. It would of course be invidious

to cite examples—in fact they would include our libraries; but the author cannot help expressing his sympathy with the universal disapprobation, almost reaching contempt, expressed for the extraordinary revelations, in the form of cross sections, made in a book published only within the year, and claiming to be a guide to American geology. If our young men are to be initiated into the mysteries of form, it must be by represented forms which at least *in some degree* agree with those in Nature. The rocks of no mountain were ever seen by man to display their section lines fanlike across its summit, as in the old, and therefore excusable sections across the Alps; or as in the new, and therefore dishonorable sections across the Taconic Mountains of Western New England. The cause of all this error is a most unfortunate prejudice which still retains its powerful hold upon us, that the human eye is not nice enough in its distinctions to be trusted with an *unexaggerated vertical scale;* it is a grand mistake. Civil engineers are obliged to exaggerate their vertical scale to save paper, and make their estimates exact down to a cubic yard. Geologists have blindly copied the practice, without the excuse of its necessity; nay, in spite of its inconveniences and mischiefs. Nothing has retarded the science of dynamic geology as this has done. Most of the innumerable cross sections which have been made of the crust, and published in the numerous proceedings of societies, and in private works, will have to be redrawn upon an equal horizontal and vertical scale, to bring their observed dips down to the true angles, and develop the curves of upheaval, and planes of non-conformability, before the theory of terrestrial structure can be finally adjusted to the truth. It will certainly surprise many an old observer, who has contented himself with distorted diagrams, to find how nicely his fingers can make out upon the paper, and his eye interpret afterwards, all the details of the most variable and gentle slope, its water-beds, steeps, terraces, and summits, when they are drawn to a precisely equal scale with the distances in horizontal line. *Then,* when he puts in his dips at the exact localities where he has observed them, he will find them flow into each other easily through curves of flexure at once both beautiful and true. His geology grows under his eye without his bidding, like the plant in his garden, by forces within itself, over which his prejudices can exert no mischievous control, and all that he has to do is to watch the development of a local generalization, for which he has the sufficient honor of having prepared the proper conditions, and thereby secured the truth.

---

42. If this manual has answered at all the design of the writer there remains little to add in the way of general direction for the prosecution of a survey. It does not appeal to absolute ignorance, as an elementary form of instruction, but it appeals to knowledge already intelligently and zealously at work, as a sympathizing and experienced train of suggestion. It can only express the habitual thinking of the hour of work, when innumerable minor exigencies call for practical answers to practical questions. These answers, however, after all, however imperfectly made, cover all the ground. The rules that govern large surveys are, after all, deduced from the adventures which befall the conduct of small surveys. To survey a State is a more onerous, a longer, a more expensive work than to survey a field, but it must proceed in precisely the same way, and by the same kind of means and men. It is only in its administrative department that a difference would be found; provision must be made for the harmonious working of many minds, and the adjustment of separate individual rights. But there are no essential difficulties to be overcome other than those which so frequently blast the fruits of a geologist's labors within the limits of a township.

43. If for instance the State of Ohio were to order a complete geological survey of its territory, the mode of procedure which the experience of the past and of neighboring States would suggest might be stated somewhat in this form.

a. The **democratic element** would be introduced as

largely as possible into the organization of the survey, so as to insure the greatest amount of personal responsibility in every member of it to the public and the world at large, intensifying his personal zeal by insuring him the reward of reputation. The example of States upon the borders of Ohio would suffice to show the different results to be expected from the two opposite systems; in one of which, men, sharing alike the responsibilities of the work, and free each one upon his own ground, separately report their results in full, each in his best manner and at the earliest day, feeding science and satisfying the demands of the times; in the other, one man invested with absolute power over the rest without a previous trial of his administrative talents, falling helplessly under the temptation to absorb, appropriate to his own ambition and suppress the greater portion of the results, delaying their publication to suit private ends, and robbing both those who do the work of their just fame and society of its just expectations. It is of prime importance that men of science should work under the excitement of a laudable ambition, and be made enterprising and circumspect by its rewards and punishments. Therefore,

b. No **primary report** should receive the touch of any hand but that of the first observer. Let it come fresh and clean before the examination of the world, and stand or fall (with its author) by its own merits. What we want is *facts at first hand;* not facts compiled, distorted, suppressed, or exaggerated to suit the theories or universalize the fame of some chief of a survey.

c. At the same time the crudity involved in an *early* publication of a primary report, and the narrowness of view and consequent inaccuracy of generalization incident to a *district* report, makes it needful at some subsequent date, and after all the primary reports are in the hands of the people and in daily use, that they be subjected to the learning and genius of some one mind, by which the general laws may be evolved and elucidated, errors corrected, discrepancies composed, and the whole work unified and harmonized with all that is real science at the time. This should be the business of some one member of the survey selected for the purpose. His responsibility would be confined to doing this work well. He would neither be responsible for blunders, nor unjustly claim merit for discoveries other than his own, for those of others would have been already published. This **final report,** including as it must distinct departments, should carry upon its title-page the several names of those who would stand responsible for its parts.

d. To obtain the utmost possible knowledge, interest, and accuracy, the **distribution of the work** would have to be regulated in such a manner as to give each member of the corps such ground as would not keep his attention constantly distracted by an infinite number of heterogeneous things, but rather permit one great subject or class of subjects to claim the principal share of his thoughts, to make him entirely master of that department, and feel in the end that he actually has something to say about it which is at least well considered and true, if not very extraordinary or very new. There should be a man for the coal, and another for the iron ore, a third for the sandstone districts, and a fourth for the lower limestones, a fifth for the drift and alluvium, a sixth for the fossil plants, a seventh for fossil shells, and an eighth for topography. Certainly no one of these would blind himself to facts coming properly under the consideration of the others; on the contrary, would become their assistants in the collection of such facts, so far as it would not interfere with his own work. But on this very account it would be necessary for the corps to be constrained by the organic law of the survey to frequent **conventions,** and a more or less permanent winter rendezvous, during which an amicable interchange of opinions and facts would go on, giving spirit and efficiency to the work. This would avoid the principal rock upon which other surveys have split, where the subordinates were so separated as to be kept in ignorance of each other's discoveries and views. If we take Ohio as a subject for these remarks, it is evident that it might be divided among a corps of geologists either lengthwise or crosswise; but the difference to the results would be immense. Any belt selected as running from east to west, crosses all the measures from the coal to the Trenton limestone ; the investigator of such a belt would learn a little of everything and not much of anything. On the contrary, the coal, the iron, the sandstones, and the lime, run north and south in narrow parallel belts, and a corps of geologists arranged abreast of each would sweep the State from north to south (or from south to north according to the seasons), each on his own ground, and all in easy communication with one another.

e. They should, however, be *preceded by a topographic corps.* This is an all important preparation. The survey of Pennsylvania would have been done in half the time, and twice as well, it is not too much to say, had we had in our hands such county maps as have been made and published by J. Pearsal Smith, of Philadelphia, of the counties of New York and other States.

Where a geologist finds himself compelled to map as he goes, his strength is exhausted, his attention distracted, and the results are after all so imperfect, that they do little more than demonstrate the necessity of a new topographical survey of the district he has so laboriously examined. This is especially true of the Appalachian region, but is true of every country. With an accurate road and water map of the district in his hand, the geologist can see where to go, can put down his notes at once in place, can make his calculations and estimates in the field, trace and identify his outcrops, and return home ready to make out his report, harmonize his results with others, and begin at once the study of his fossils or the analysis of his ores. But without this preliminary map, the winter drags itself along under his weary fingers in innumerable poor plottings and conjectural compilations, and the consequence is a mass of results without certain relationships to one another, full of lacunæ, which he could not be aware of on the ground, and imperfectly expressed at the best. Such is the history of American geology up to the present day, with one or two exceptions, and these more apparent than real. Nothing will remedy this radical evil but the reversal of our method, putting topography before geology or with it, instead of after it. The importance of the subject will justify me in going into some details.

44. The expense of a foreign ordinance survey and of our national coast survey is enormous—necessarily so—the world, however, willingly bears it; must have the work done; none but great governments can do it. There is, however, a shorter, cheaper kind of topographical survey which answers just as well for our purpose in geology. It is also expensive, but the expense is in no proportion to that of astronomical and trigonometrical surveys. It determines magnetically, with sufficient accuracy for all practical purposes, the features of the surface, and enables observers to locate with precision their facts, and deduce with ease their results. A county surveyor with a light odometer and common compass and one boy, can survey all the roads of a county of average size, say 20 miles square, and put in all the buildings, water-courses, &c. in a single season or part of one. Nearly sixty counties of New York have been mapped thus by Mr. Smith, for about thirty thousand dollars, by letting the work, at various bids, to county surveyors, students, and amateurs. The results are perfectly satisfactory. A topographer with a good assistant, and a pair of the best aneroid barometers, could take any county map so constructed, and draw in upon it all the contour lines of its surface in a fortnight. If a geologist, he could lay in at the same time, only a little more slowly, all the outcrops of the internal structure. After this is done, nothing remains but the economic, analytic, and palæontological study of the specimens.

45. It would be a great oversight, however, and poor economy, when these county maps are to be compiled into one State map, not to perfect this practical system of surveys and lift it nearer to the grade of accurate science, by the astronomical determination of a score or two of points, principal cities, county-seats,

colleges, &c., from which all the practical surveys would begin their plotting. This would cost from fifteen to twenty thousand dollars more, but would remunerate the State by its effect upon the business in the land-title office.

If now we imagine a well arranged corps of geologists beginning their work in a line across the State with maps in their hands constructed the previous season, of as many counties as they would be likely to examine, we may imagine county surveys at the same time going on in advance of them, covering a second belt of the State which they are to attack the season following. The season following, while they examined these newly surveyed counties, the counties still in advance would be surveying; and so on to the end. Thus, the work would distribute itself equally over the shortest desirable time, and the principal expense—that of topography—would not be felt so onerous. Such an organization should survey the whole State of Ohio in three years, or at the outside four, and give all the practical results complete to the public, from year to year, as it went on; reserving for the end only those deductions of science, the only use of which to the public comes through the education of its men of science and therefore are not immediately but yet to the last degree valuable.

APPENDIX.

I. p. 20.

Fireclay is a compound of silica and alumina, sometimes contains magnesia, and more often a small quantity of lime. Iron injures its quality if present in any quantity. It is used for linings for furnaces and is worth $2 00 raw, and $30 per thousand made up into bricks.

II. p. 51.

The *Permian* New Red of Russia is supposed indeed by Murchison and others to ally itself more closely with the carboniferous below, than with the new red above, but it is yet to be demonstrated that it is not an integral member of the latter.

III. p. 60.

This great question claims a volume to itself. The author is convinced by all he has seen in the south, that most of what is there called primary rock is Silurian if not Devonian metamorphosed. The long, narrow anthracite coal basin, ending northeastward at Pittsylvania Court House, in Virginia, and southwestward in Guildford Co., North Carolina—never more than four miles wide, and resting on what is called "syenite" all round, a synclinal basin containing one coal bed of 4 feet thick and several thinner seams—undoubtedly rests in the centre of a great trough of Devonian and Silurian rocks. The author has also been lately shown specimens of a porphyritic conglomerate from an outcrop which entirely surrounds the Richmond coal basin. The rocks of the Ocooee and Hiwasse Rivers in East Tennessee, traversed in going across from the Coal to the Copper mines, are

Palæozoic shales, sands and limestones metamorphosed—and that often but very slightly; that whole structure is yet to be made out, and will repay the labor of doing it.

IV. p. 65. DESCRIPTION OF FIG. 13.

	FEET.	FEET.
Limestone, 28. *Shale* 2. Low water Mississippi,	30	30
Limestone,	231	261
Chertrock, drab,	15	276
Limestone, drab,	74	350
Shale, dark green,	30	380
Limestone, 75. *Shale*, dark green, 1½,	76½	456½
Limestone, drab, 38½. *Shale* sandy, 6½,	45	501½
Limestone, dove,	128½	630
Red Marl,	15	645
Shale, dark greenish,	30	675
Red Marl,	50	725
Shale, dark green,	30	755
Limestone, olive,	119	874
Shale, dark lead,	66	940
Bituminous Marl,	15	950
Shale, dark lead,	85	1035
Limestone, olive,	134	1169
Cherty rock, dove,	62	1231
Limestone, dove, 120. *Limestone*, 18,	138	1351
Shale,	17	
Limestone,	20	
Shale,	56	
Limestone,	37	
Sand-rock, white,	15	
Sandstone?	160	1511
Sand and iron at 1636. Sand and clay,		1641
Sandstone, 1648,		1673
Sandstone and Limestone, 1792—1223, (no specimens kept.)		1223
Sand, 1823. Clay, 1829. Lime, 1392. (No specimens.)		1391
for 200 feet. Yellow sand, 2193,		2193

N. B.—[The specimens from 1600 to 1820 showed a reddish grain, mixed with white and yellow.]

N. B.—First appearance of gas, 467. Sea level, 520. Salt water got first about 600. [Color of 645 like IX, not XI. Specimen from

680 a true XI red shale triturated and dried. Spec. 835—874, powder of gray silic. lime, hard, compact, close, fine fracture.] 765. **Salt water**, $1\frac{7}{8}$ p. c. salt. 919—932. Brown clay, alum, silic. lime, including coaly matter and iron; Dec. 29, 1851; [looks like coal crop in road loam.] 932—1002. Blue slate, alum, sil. lime, and iron, [for 80 feet troublesome boring; tubed.] **Salt water**, 1015, $2\frac{1}{4}$ p. c. salt; 1211, 3 p. c. salt. 1236, magn. lime; 1244, bottom 69°, top, 50° Fahrenheit, 1270, much less gas; 1280, much gas; 1301, less gas. 1343, bottom 65° F., top, 50° F.; March 19. 1360, Magnes. lime, strong sulph. hyd. 1409 gas, and saltish water. 1472 magnes. lime. 1509, strong sulph. hyd. gas. 1532, [specimen, pure, white, silic. sand, like the purest X or XII.]

V. p. 82.

Description of Sections.

VI. p. 90.

Fossil Plants of the Coal.—A very good list, but made up only to 1836, is to be found in *Thompson's Outlines of Mineralogy, Geology, and Mineral Analysis.* (London, vol. ii., p. 265.)

I. DICOTYLEDONS.

Genus		Species.	Genus		Species.
I.	**Sigellaria,**	33	IX.	**Asterophyllites,**	13
II.	**Favularia,**	1	X.	**Pinnularia,**	1
III.	**Stigmaria,**	8	XI.	**Hippurites,**	1
IV.	**Bothrodendron,**	1	XII.	**Megaphyton,**	2
V.	**Pinites,**	4	XIII.	**Halonia,**	2
VI.	**Knorria,**	2	XIV.	**Phyllotheca,**	1
VII.	**Sphenophyllum,**	7	XV.	**Annularia,**	7
VIII.	**Peuce,**	1	XVI.	**Bechera,**	1

II. MONOCOTYLEDONS.

Genus		Species.	Genus		Species.
I.	**Nœggerathia,**	2	IV.	**Cyperites,**	1
II.	**Flabellaria,**	1	V.	**Poacites,**	2
III.	**Cannophyllites,**	1	VI.	**Sternbergia,**	3

III. FILICES.

Genus	Species.	Genus	Species.
I. **Cyclopteris,**	5	V. **Sphenopteris,**	39
II. **Glossopteris,**	2	VI. **Neuropteris,**	25
III. **Schizopteris,**	2	VII. **Pecopteris,**	45
IV. **Caulopteris,**	1	VIII. **Odontopteris,**	5

IV. LYCOPODIACEÆ.

Genus	Species.	Genus	Species.
I. **Lycopodites,**	8	IV. **Lepidophyllum,**	1
II. **Selaginites,**	2	V. **Lepidodendron,**	42
III. **Ulodendron,**	2		

V. EQUISETACEÆ.

Genus	Species.	Genus	Species.
I. **Equisetum,**	2	II. **Calamites,**	13

VI. CONFERVACEÆ.

Genus I. **Confervites,** 1 species.

VII. FUCACEÆ.

Genus I. **Fucoides,** 1 specimen.

VII. p. 138.

The equivalents of the old Pennsylvania and Virginia nomenclature are not yet absolutely fixed in the New York, English and Continental systems, but within certain limits may be represented thus. See among others Hall's Parallelism, &c. in vol. ii. chap. xviii. of Foster and Whitney's Lake Superior Report.

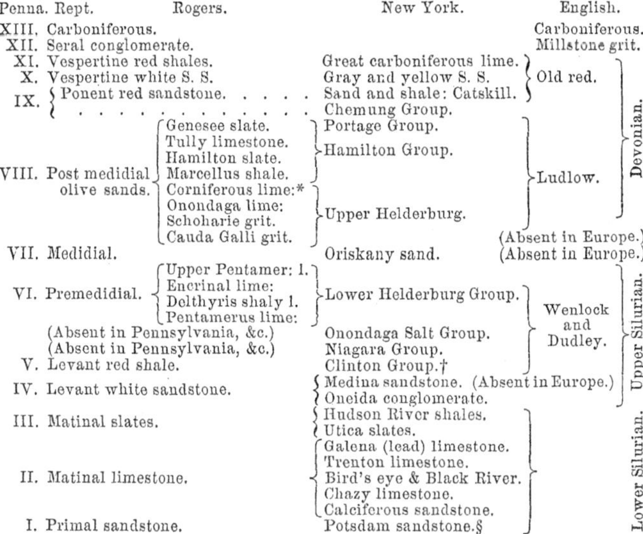

Penna. Rept.	Rogers.		New York.	English.	
XIII.	Carboniferous.			Carboniferous.	
XII.	Seral conglomerate.			Millstone grit.	
XI.	Vespertine red shales.		Great carboniferous lime.	Old red.	Devonian.
X.	Vespertine white S. S.		Gray and yellow S. S.		
IX.	Ponent red sandstone.		Sand and shale: Catskill.		
	. . . .		Chemung Group.		
VIII.	Post medidial olive sands.	Genesee slate.	Portage Group.	Ludlow.	
		Tully limestone. Hamilton slate. Marcellus shale.	Hamilton Group.		
		Corniferous lime:* Onondaga lime: Schoharie grit. Cauda Galli grit.	Upper Helderburg.	(Absent in Europe.)	
VII.	Medidial.		Oriskany sand.	(Absent in Europe.)	
VI.	Premedidial.	Upper Pentamer: l. Encrinal lime: Delthyris shaly l. Pentamerus lime:	Lower Helderburg Group.	Wenlock and Dudley.	Upper Silurian.
	(Absent in Pennsylvania, &c.)		Onondaga Salt Group.		
	(Absent in Pennsylvania, &c.)		Niagara Group.		
V.	Levant red shale.		Clinton Group.†		
IV.	Levant white sandstone.		Medina sandstone. Oneida conglomerate.	(Absent in Europe.)	
III.	Matinal slates.		Hudson River shales. Utica slates.		Lower Silurian.
II.	Matinal limestone.		Galena (lead) limestone. Trenton limestone. Bird's eye & Black River. Chazy limestone. Calciferous sandstone.		
I.	Primal sandstone.		Potsdam sandstone.§		

* Black slates of Ohio, Indiana, and Kentucky come in over corniferous limestone. The Onondaga and corniferous limestones Verneuil makes equivalents of the lower rocks of the Eifel and Hartz; the Oriskany sandstone equivalent of the Rhine fossil schists.

† The Clinton Group is the upper Caradoc sandstone of England.

‡ Llandello flags and Caradoc sandstone of Britain; Orthoceratite beds of Russia and Sweden.

§ Lingula beds of Russia.

VIII. p. 146.

The head of Mt. Washington for about 400 feet down differs from all the rest of the White Mountain range in being covered with a sheet of oblong, curved, angular fragments of altered sand-rock. The slopes below are rounded, polished, and grooved, with boulders lying upon them. This head has escaped the denuding action, and is the only fact of any magnitude which looks like glacial action there. The summit of the Peaks of Otter, in Virginia, is similarly broken, and is said to consist of a "magnetic granite," each fragment of which, however small, exhibits double polarity.

IX. p. 161.

One of the finest cases of this cleavage structure topography is to be seen in the Jahr-buch: Kaiser: Kœnig's Geologischen Reichenstadt Wien, for 1855. No. 1, p. 111; a picture of the double peaks of Petuli Domni, two hundred feet high.

X. p. 162.

Prof. Rogers has beautifully applied his theory of thermoelectric anticlinal axis lamination to explain the **ribbon structure** of glaciers, which turn down the edges of their bands against the walls of the valley, and up against the open air at the end of the glacier, thus presenting a vertical or transverse crystallization to the planes of different temperature.

XI. p. 174.

Prof. Hall explains the turning away of the river at the whirlpool *from* the diluvium of St. David's valley, *to* the solid rock through which it cuts to Queenstown, by reference to similar phenomena elsewhere in New York. For instance, the Genesee River near the Portage Falls leaves a wide valley full of diluvium, and cuts its way across through solid rock. But while I accept the facts, I cannot persuade myself that a mighty river would find it more difficult to sweep off gravel than to cut through rock. The cutting must have been done by another agency. But in the case of the Niagara River, it is not a question of *difficulty*, but of impossibility; for the elevation of the brow of the escarpment at Brock's Monument, where the falls are supposed to have commenced their cutting, must be so much higher than the St. David's valley that the river could have had no alternative, but must have scoured out the diluvium. Fig. 66 is designed to show the similar structure of Niagara Falls, A, the Sawkill Falls, Milford, Pa., B. The Towanda Falls, Bradford Co., Pa., C, and the side slopes of the Pocono Mountain, Upper Silurian and Devonian, Pa., D.

Fig. 66.

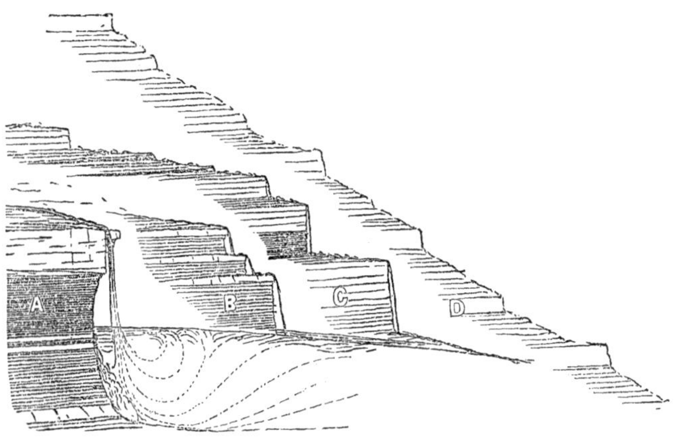

XII. p. 178.

P. W. Sheafer's **estimates of the anthracite coal trade**: 1830, 174.734 tons ; 1840, 865.414 ; 1850, 3,356,899 ; 1860, 10,000,000 tons.

Comparison of Yield of North and South Dipping Veins.

North dip,	10 collieries,	red ash,	84,732 ;	5	white ash,	91,222 tons.
South "	48	"	" 570,561 ;	26	"	745,231 "
N. and S. "	11	"	" 305,022 ;	5	"	120,101 "

The north dips are steeper in the Pottsville basin than the south, and therefore more slipped and crushed, thinner, and more broken. This is one of the principal arguments for the Wave Theory of Rogers.

XIII. p. 47.

Sediment of the Mississippi. (See Proceedings of Philad. Meeting Amer. Assoc., 1848.)

Annual Aggregate Water (a), (diminishing), 15,000,000,000,000 cubic feet, or 101 cubic miles. Sediment : water : : 1 : 528.

Annual Aggregate Sediment, 28,000,000,000 cubic feet.

Delta, estimated 13,600 sq. miles, 1056 ft. deep (aver. depth of gulf), = 400,000 000,000 000 cub. feet, = 2720 cub. miles, = 14,203 years at present rates.

Valley, 1,400,000 sq. miles; fall of rain 52 in. (at Natchez 55), = 169,000 000,000 000 cub. feet, = 12 × a; 11 × a evaporates. Flow past New Orleans in 1823 : 1848 : : 100 : 75–80. Hence, the bottom lands are redeemed; fogs disappear; only half the timber comes down.

NEW SPECIMENS OF AMERICAN FOSSIL PLANTS.

BY LEO LESQUEREUX.

Genus.	Species.	Genus.	Species.
Calamites. Brongt.	2	Pecopteris. Brongt.	7
Asterophyllites. Brongt.	5	Crematopteris. W. P. Schimper.	1
Annularia. Sternb.	1	Scolopendrites. Lesqx.	1
Sphenophyllum. Brongt.	2	Caulopteris. Lindl. & Hutt.	2
Nœggrathia. Sternb.	3	Stigmaria.	5
Cyclopteris.	5	Sigillaria.	9
Neuropteris.	13	Lepidodendron. Sternb.	10
Odontopteris. Brongt.	2	Lepidophyllum. Brongt.	6
Sphenopteris. Brongt.	8	Brachyphyllum? Brongt.	1
Hymenophyllites. Göpp.	3	Cardiocarpon. Brongt.	3
Pachyphyllum. Lesqx.	5	Trigonocarpum. Brongt.	1
Asplenites. Göpp.	1	Rhabdocarpus. Göpp. & Berg.	1
Alethopteris. Sternb. and Göpp.	5	Carpolithes. Sternb.	3
Callipteris. Brongt.	1	Pinnularia.	5

www.ingramcontent.com/pod-product-compliance
Lightning Source LLC
LaVergne TN
LVHW050530100826
845148LV00002B/501

* 9 7 8 1 4 2 5 5 1 9 1 8 6 *